Lecture Notes in Mathematics

Edited by A. Dold and B. Eckmann

948

Functional Analysis

Proceedings of a Conference
Held at Dubrovnik, Yugoslavia,
November 2 – 14, 1981

Edited by D. Butković, H. Kraljević, and S. Kurepa

Springer-Verlag
Berlin Heidelberg New York 1982

Editors

Davor Butković
Department of Applied Mathematics, Electro-engineering Faculty
Unska 3, 41000 Zagreb, Yugoslavia

Hrvoje Kraljević
Svetozar Kurepa
Department of Mathematics, University of Zagreb
P.O. Box 187, 41001 Zagreb, Yugoslavia

AMS Subject Classifications (1980): 22E15, 22E45, 22E70, 28A35, 28B05, 46D05, 46H05, 47B50, 47D05, 47H10, 60B12, 60F05, 60F15

ISBN 3-540-11594-3 Springer-Verlag Berlin Heidelberg New York
ISBN 0-387-11594-3 Springer-Verlag New York Heidelberg Berlin

Printed in Germany

Printing and binding: Beltz Offsetdruck, Hemsbach/Bergstr.
2146/3140-543210

FOREWORD

This volume contains lecture notes given at postgraduate school and conference on Functional Analysis held from November 2 to November 14, 1981, at the Interuniversity Center of postgraduate studies, Dubrovnik, Yugoslavia.

The conference and the school were devoted to several parts of functional analysis but centered mainly on operator theory in Hilbert spaces. The lectures were divided in five areas:

1. Operator theory on Hilbert spaces, given by P.R.Halmos (Indiana University, Bloomington);
2. Hilbert spaces with an indefinite metric, given by H.Langer (Technische Universität, Dresden);
3. Semigroups and cosine operator functions, given by S.Kurepa (University of Zagreb) and by D.Lutz (Universität Essen, Gesamthochschule);
4. Analysis on groups, given by A. Guichardet (Ecole Polytechnique, Palaiseau);
5. Geometry of Banach spaces and probability, given by J.Hoffmann-Jørgensen (Aarhus Universitet).

Besides these topics, there were some other one -to two -hour lectures and also shorter contributions and communications by attendants.

Professor P.R.Halmos has given six lectures on operator theory in Hilbert spaces. Many problems (some easy, some more difficult) have been described. The problem of approximating an arbitrary bounded operator by: selfadjoint operators, positive operators, unitary operators, scalar operators, normal operators etc. has been considered. Furthermore,the problem of approximating a given operator by an operator with prescribed spectrum has been studied. In connection with approximation problems the (strong) closure of the set of all: projections, unitary operators, isometries, co-isometries, normal operators, hyponormal operators etc. was considered. Finally, problems related to the essential spectrum, the essential commutant of the unilateral shift U (i.e. the question for which T is $T - U^*TU$ compact), the continuity properties of the function $A \to \text{lat } A$ (associating to each operator A its lattice of all invariant subspaces) and some other problems were investigated.

These lectures have been already published in one or another form and, therefore, they are not included in these proceedings.

In 1944 L.S.Pontrjagin investigated spaces with indefinite inner product and selfadjoint operators in them. These investigations were afterwards continued by M.G.Krein, I.S.Iohvidov and others. Two most important types of such spaces are called Pontrjagin and Krein spaces. The main questions concerning selfadjoint operators in Krein spaces are the existence of maximal nonnegative invariant subspaces and of spectral functions of definitizable operators. It should be noted that the existence of the spectral function is helpful in solving the nonnegative invariant subspace problem. The existence and the properties of the spectral function of a selfadjoint operator were first proved by M.G. Krein and H.Langer in the case of Pontrjagin spaces. Later H.Langer extended these results to the general case of definitizable operators in Krein spaces. His results appeared in his Ph.D. thesis and were often quoted without proofs. In his lecture in this volume H.Langer publishes all the original proofs of these, nowadays classical, results.

Exponential and cosine functions are closely related to differential equations

$$x'(t) = Ax(t) \quad , \quad x''(t) + Ax(t) = 0 \quad ,$$

with corresponding initial conditions, and with the following functional equations:

$$E(t+s) = E(t)\,E(s), \quad E(0) = 1 \ , \ t,s \in R \ ,$$

$$C(t+s)+C(t-s)=2C(t)C(s), \quad C(0) = 1 \ , \ t,s \in R \ .$$

It has been observed by A.Cauchy that the above functional equations are characteristic for exponential and cosine functions. This fact lies behind the idea to define exponential and cosine functions in abstract structures as functions which satisfy the above functional equations and, of course, some kind of regularity conditions. The theory developed along this idea for exponential functions E and differential equation $x'=Ax$ is well-known as the theory of semigroups. The parallel theory for cosine functions C and differential equation $x''+Ax = 0$ is known as the theory of abstract cosine functions or cosine operator functions. The theory of semigroups was investigated in detail in many monographs, and lectures by S.Kurepa and D.Lutz were concerned mostly with abstract cosine functions. Questions concerning regularity of cosine functions, cosine functions in Banach algebras and cosine functions with values in the set of bounded normal operators were treated in the lectures by S.Kurepa and questions concerning the differential equation $x''+Ax = 0$, infinitesimal generators of cosine functions, Hille-Yoshida-

-type criterion for the infinitesimal generator of a cosine function, perturbations of a generator etc. were investigated in the lectures by D.Lutz.

The central problem in the representation theory of locally compact groups is the following: given a locally compact group G describe the set $\hat{G}$ of all equivalence classes of irreducible unitary representations. The theory of induced representations, developed by G.W.Mackey in fifties, allows to reduce it to the description of $\hat{H}$ for some proper subgroups H of G , provided G possesses a non-central closed normal subgroup. In his lectures A.Guichardet explained basic notions and fundamental results in the theory of Lie groups and Lie algebras and in the repreesentation theory of Lie groups. The representation theory of compact Lie groups is illustrated in the easiest example of the group SU(2). Furthermore, the Mackey theory of induced representations is described and applied to the case of Galilean group G (the invariance group of classical mechanics) to obtain the description of $\hat{G}$.

Probability in Banach spaces is now almost thirty years old; one considers that it originated in 1953 in the thesis by Edith Mourier, where the first Strong Law of Large Numbers for random variables with values is Banach spaces was proved. As a subject of study, such a theory is motivated by the theory of stochastic processes, where we can represent a process as a random element in some space of functions. In the later development it was more and more clear that the truth or falsity of probabilistic theorems in Banach spaces are closely related to the geometry of the space. Thus the study of probability theorems introduces some classes of Banach spaces and these two subjects are connected in a rather surprising manner. The lectures by J.Hoffmann-Jørgensen are mostly concerned with this relationship, especially connected with Laws of Large Numbers and Central Limit Theorems. The exposition of the subject follows an alternative way with respect to the one taken by the same author in his lectures in Ecole d'Eté de Probabilités de Saint-Flour VI- 1976 (whose proceedings were published in Springer-Verlag Lecture Notes N° 598).

Besides these five lecture notes the volume contains four papers connected with some one to two-hour lectures.

We use this opportunity to express our thanks to the institutions whose financial support made the conference possible. These are: Samoupravna interesna zajednica za znanstveni rad SRH - SIZ VI, Department of Mathematics , University of Zagreb, Departments of Applied Mathematics of the Building Faculty and of the Electroengineering Faculty. We are also grateful to the Interuniversity Center of postgraduate studies in Dubrovnik where the postgraduate school and the conference were hold.

All the typing was done by Mrs Božena Grdović; the reader should appreciate the quality of her work and we thank her for the efficiency and patience.

Although all the papers were proofread by the authors, the editors bear responsibility for any inaccuracies they contain, especially because some corrections were not done due to the pressures of time; the editors would also be grateful to have the authors' understanding in this matter.

Finally, the editors are grateful to Springer-Verlag for its prompt publication of these proceedings.

S.Kurepa
H.Kraljević
D.Butković

CONTENTS

ADDRESSES OF THE AUTHORS

Davor BUTKOVIĆ,	Department of Applied Mathematics, Electroengineering Faculty, Unska 3, 411000 Zagreb, Yugoslavia
Alain GUICHARDET,	École Polytechnique, Centre des Mathématiques, Plateau de Palaiseau, Palaiseau, France
Olga HADŽIĆ,	Faculty of Mathematics and Natural Sciences, Jovana Subotića 7, 21000 Novi Sad, Yugoslavia
Jørgen HOFFMANN-JØRGENSEN,	Aarhus Universitet, Matematisk Institut Ny Munkegade, DK-800,Aarhus C, Denmark
Hrvoje KRALJEVIĆ,	Department of Mathematics,University of Zagreb, Marulićev trg 19, Zagreb,Yugoslavia
Svetozar KUREPA,	Department of Mathematics, University of Zagreb, Marulićev trg 19, Zagreb,Yugoslavia
Heinz LANGER,	Technische Universität Dresden, Sektion Mathematik, Mommsenstrasse 13, 8027 Dresden, DDR
Dieter LUTZ,	Gesamthochschule, FB 6 Mathematik, Universitätstr. 3, 4300 Essen 1, BRD
Aljoša VOLČIČ,	Istituto di Matematica applicata, Universitá di Trieste, Piazzale Europa 1, 341000 Trieste, Italia

SPECTRAL FUNCTIONS OF DEFINITIZABLE OPERATORS IN KREIN SPACES

Heinz Langer

INTRODUCTION

In 6 lectures on "Hilbert spaces with an indefinite metric" an introduction to the theory of linear operators in Krein spaces and their applications was given. The topics of the lectures were the following:

1. Definitions. Examples. Geometry.
2. Maximal nonnegative invariant subspaces.
3. Spectral functions of definitizable operators.
4. Some classes of analytic functions.
5. Sturm-Liouville operators with an indefinite weight function.

The present notes contain a somewhat extended version of the parts 1 and 3 of these lectures. The results of the parts 2,4 and 5 can be found in the literature: For part 2 we refer to J.Bognár's book [1] and to T.Ando's lecture notes [2] , for part 4 to [3] and the literature quoted there, the results of part 5 can be found in [4].

Thus, the main topic of these notes is the spectral function of a definitizable operator in a Krein space. These results were originally obtained in [5] but the complete proofs given there have not yet been published. For the special case of a bounded nonnegative operator in a Krein space the existence of a spectral function was shown by M.G.Krein and Ju.L.Šmul'jan in [6] and by T.Ando in [2] , for an arbitrary definitizable operator by P.Jonas [7] . Although Jonas' proof, based on a functional calculus, is shorter and Ando's proof (for the special case) is more elegant, we think that it still may be of some interest to publish the original proof from [5] . It uses Cauchy principal value integrals of the resovlent which are a classical tool in the spectral theory in Hilbert and Banach spaces and, in particular, in its applications (e.g. in the spectral theory of differential operators). The main ideas of this proof are simple, the more technical results in II.1 can be considered to be well-known.

Moreover, by means of these resolvent integrals it is easy to prove some perturbation results, e.g. extensions of Rellich's theorem about the convergence of the spectral function, see [8].

In sections II.4-6 we study the critical points of definitizable and, in particular, of nonnegative operators. These results are partially taken from [5] and published here the first time. In section II.7, as an application of the spectral function it is shown that a nonnegative bounded operator in a Krein space has a maximal nonnegative invariant subspace. More general results of this type can be found in [9] .

It was also an aim of the lectures to show that nowadays there is a variety of problems in which results on operators in Krein space and, in particular, on definitizable operators play an essential role. For this reason in section I.3 we list some examples. However, neither this list nor the topics mentioned at the beginning of this introduction give a complete information. The reader, interested in other aspects of the theory on spaces with indefinite metric and its applications, should consult the book [1] , the lecture notes [10],[2] and the survey article [11](containing more than 380 references!). There he may also find historical remarks which are avoided here.

The other sections of chapter I contain the necessary definitions and a minimum of geometry and elementary spectral properties of selfadjoint operators in Krein spaces which is necessary for chapter II. Thus these notes are essentially selfcontained. However, some knowledge of the results on Pontrjagin spaces (see,e.g. [12] and [1]) will be useful. We also mention that for Pontrjagin spaces the spectral functions and their critical points have much more special properties than those given in chapter II, see [13] ,[5].

In these notes we restrict ourselves to selfadjoint operators. However, the existence of a spectral function for a definitizable unitary operator can be proven in the same way. In [5] this was done first and the results for selfadjoint operators were obtained by means of the Cayley transformation.

I. DEFINITIONS AND EXAMPLES

I.1. Krein spaces. Let K be a complex linear space. A scalar product (s.p.) on K is a nondegenerated Hermitian sesquilinear form on K, that is a mapping $[\cdot,\cdot] : K \times K \to C$ with the following properties:

(i) $[\alpha_1 x_1 + \alpha_2 x_2, y] = \alpha_1 [x_1, y] + \alpha_2 [x_2, y]$ $(x_1, x_2, y \in K,\ \alpha_1, \alpha_2 \in C)$.

(ii) $[x,y] = \overline{[y,x]}$ $(x, y \in K)$.

(iii) $[x_o, y] = 0$ for some $x_o \in K$ and all $y \in K$ implies $x_o = 0$.

If the space K is equipped with the s.p. $[\cdot,\cdot]$, it will often be denoted by $(K, [\cdot,\cdot])$. The s.p. will sometimes be called indefinite, if there exist elements $x, y \in K$ such that $[x,x] > 0$, $[y,y] < 0$.

The space $(K, [\cdot,\cdot])$ is called a Krein space, if it contains two subspaces K_+, K_- with the properites:

1) $K = K_+ \dot{+} K_-$, (1.1)

2) $(K_+, [\cdot,\cdot])$ and $(K_-, -[\cdot,\cdot])$ are Hilbert spaces,

3) $[K_+, K_-] = \{0\}$

If, in particular, $\kappa := \min(\dim K_+, \dim K_-) < \infty$, the Krein space $(K, [\cdot,\cdot])$ is called a Pontrjagin space of index κ or a π_κ-space. In this case $\dim K_+$ ($\dim K_-$, respectively) is called the number of positive (negative, respectively) squares of the scalar product $[\cdot,\cdot]$.

The condition 2) means that, e.g. on K_+ the s.p. $[\cdot,\cdot]$ is positive definite: $[x,x] > 0$ if $x \in K_+$, $x \neq 0$, and that K_+ is complete with respect to the norm $[x,x]^{1/2}$ $(x \in K_+)$. The possibility that one of the spaces K_+, K_- consists only of 0 is not excluded.

Using the decomposition (1.1) a positive definite s.p. $(\cdot,\cdot)$ can be defined on K as follows:

$$(x,y) := [x_+, y_+] - [x_-, y_-] \quad (x = x_+ + x_-,\ y = y_+ + y_-,\ x_\pm, y_\pm \in K_\pm) \quad (1.2)$$

It is easy to see that $(K, (\cdot,\cdot))$ is a Hilbert space. Moreover, $(K, (\cdot,\cdot))$ is the orthogonal sum of the Hilbert spaces $(K_+, [\cdot,\cdot])$ and $(K_-, -[\cdot,\cdot])$. Introducing the projectors $P_\pm$:

$$P_{\pm}x: = x_{\pm} \text{ if } x \in K,\ x = x_+ + x_-,\ x_{\pm} \in K_{\pm} \quad ,$$

and $J: = P_+ - P_-$, we have $J^2 = I, J = J^*$ (here $*$ denotes the adjoint with respect to the s.p. $(\cdot,\cdot)$) and

$$[x,y] = (Jx,y),\ (x,y) = [Jx,y] \qquad (x,y \in K)\ . \tag{1.3}$$

The s.p. $(\cdot,\cdot)$ depends on the decomposition (1.1) which is, in general, not unique. However, for two such decompositions (1.1) and

$$K = K'_+ \dotplus K'_-$$

the dimensions of the corresponding components coincide:

$$\dim K_{\pm} = \dim K'_{\pm} \quad ,$$

and the positive definite s.p.-s $(\cdot,\cdot)$ and $(\cdot,\cdot)'$ generate equivalent norms. These facts are easy consequences of the Propositions 1.1 and 1.2 below.

We put $\|x\| : = (x,x)^{1/2}$ $(x \in K)$. All the topological notions in K, if no other topology is explicitely mentioned, are to be understood with respect to this Hilbert space norm topology.

If K is a linear space with a Hermitian sesquilinear form $[\cdot,\cdot]$, that is (i) and (ii) hold, we define

$$P_+ : = \{x: [x,x] \geq 0,\ x \in K\},$$

$$P_- : = \{x: [x,x] \leq 0,\ x \in K\},$$

$$P_o : = P_+ \cap P_- \ .$$

The elements of $P_+(P_-, P_o)$ are called <u>nonnegative</u> (<u>nonpositive</u>, <u>neutral</u>, respectively). Moreover, $x \in K$ is called <u>positive</u> (<u>negative</u>) if $[x,x] > 0$ (< 0, respectively). If all the nonzero elements of a linear manifold $L \subset K$ are positive (nonnegative etc.), then L is called <u>positive</u> (<u>nonnegative</u> etc. respectively). If K is a Krein space, the linear manifold L is called <u>uniformly positive</u> (<u>uniformly negative</u>) if

$$[x,x] \geq \gamma\|x\|^2 \quad (\leq -\gamma\|x\|^2 \text{ , respectively}) \qquad (x \in L)$$

for some $\gamma > 0$. A subspace of a Krein space means always a closed linear manifold. In the rest of this section the subspaces K_+ and

K_- are always equipped with the Hilbert s.p. $[\cdot,\cdot]$ and $-[\cdot,\cdot]$, respectively, and the corresponding norms.

PROPOSITION 1.1. <u>If L is a nonnegative subspace of the Krein space K there exists an orthogonal projector P_+^L in the Hilbert space K_+ and a contraction K^L from $P_+^L K_+$ into K_- such that</u>

$$L = \{ x_+ + K^L x_+ : x_+ \in P_+^L K \}.$$

<u>The subspace L is positive (uniformly positive, respectively) if and only if $\|K^L x_+\| < \|x_+\|$ for all $x_+ \in P_+^L K$ ($\|K^L\| < 1$, respectively).</u>

PROOF. If $x \in L$ we have $\|P_- x\| \leq \|P_+ x\|$, hence

$$\|P_+ x\|^2 \leq \|x\|^2 = \|P_+ x\|^2 + \|P_- x\|^2 \leq 2\|P_+ x\|^2 .$$

Therefore the operator $P_+|_L$ has a bounded inverse F defined on $P_+ L$: $Fx_+ = x$ if $x \in L, P_+ x = x_+$. Denoting the orthogonal projector in K_+ onto $P_+ L$ by P_+^L and

$$K^L P_+ x := P_- F P_+ x , \qquad (x \in L)$$

the representation (1.4) follows. The other statements in Proposition 1.1 are now easy to check.

A subspace $L \subset P_+$ is called <u>maximal nonnegative</u> if it is not properly contained in another nonnegative subspace.

COROLLARY 1. <u>The subspace $L \subset P_+$ is maximal nonnegative if and only if $P_+ L = K_+$. Each nonnegative subspace is contained in a maximal one.</u>

Indeed, $L \subset P_+$ implies $L \dot{+} (K_+ \ominus P_+ L) \subset P_+$. Therefore, if $K_+ \neq P_+ L$, then L cannot be maximal nonnegative. On the other hand, if $P_+ L = K_+$ and $L \subsetneqq L_1 \subset P_+$, then $K_+ \supset P_+ L_1 \supsetneqq P_+ L = K_+$, a contradiction.

COROLLARY 2. <u>All the maximal nonnegative subspaces of the Krein space K have the same dimension</u> (as K_+).

The relation (1.3) implies that the s.p. $[\cdot,\cdot]$ is continuous with respect to the norm $\|x\|$:

$$|[x,y]| \leq \|x\| \cdot \|y\| . \qquad (x,y \in K) .$$

PROPOSITION 1.2. <u>Any two Banach space norms on $(\mathcal{K},[\cdot,\cdot])$, such that the s.p. $[\cdot,\cdot]$ is continuous (in one or, equivalently, in both variables) with respect to each of these norms, are equivalent</u>.

PROOF. Let $\|\cdot\|$ and $\|\cdot\|'$ be two such Banach space norms on $\mathcal{K}$ and consider the norm

$$\|x\|'' := \|x\| + \|x\|' \qquad (x \in \mathcal{K}) .$$

Then $\|\cdot\|''$ is also a Banach space norm on $\mathcal{K}$. Indeed, given a sequence $(x_n) \subset \mathcal{K}$ such that $\|x_n - x_m\|'' \to 0$ $(n,m \to \infty)$, it follows that there exist elements x_o, x_o' such that $\|x_n - x_o\| \to 0$ and $\|x_n - x_o'\|' \to 0$ $(n \to \infty)$. Therefore $[x_n - x_o, y] \to 0$ $(n \to \infty)$ for all $y \in \mathcal{K}$ and the same relation holds with x_o replaced by x_o'. This yields $[x_o - x_o', y] = 0$ for all $y \in \mathcal{K}$, hence $x_o = x_o'$ and $\|x_n - x_o\|'' \to 0$ $(n \to \infty)$ follows. Evidently, $\|x\| \leqq \|x\|''$ $(x \in \mathcal{K})$. According to a theorem of Banach the norms $\|\cdot\|$ and $\|\cdot\|''$ are equivalent on $\mathcal{K}$. As the same holds for $\|\cdot\|'$ instead of $\|\cdot\|$, the statement follows.

I.2. <u>Typical situations</u>. Often Krein spaces arise as follows:

(a) Let $(\mathcal{K},(\cdot,\cdot)_o)$ be a Hilbert space and G a bounded and boundedly invertible indefinite selfadjoint operator in $(\mathcal{K},(\cdot,\cdot)_o)$. Define on $\mathcal{K}$ an s.p. by the relation

$$[x,y] := (Gx,y)_o \qquad (x,y \in \mathcal{K}).$$

Then $(\mathcal{K},[\cdot,\cdot])$ is a Krein space.

Indeed, we can choose, e.g.,

$$\mathcal{K}_- := E_o \mathcal{K}, \quad \mathcal{K}_+ := (I - E_o)\mathcal{K} ,$$

where E_λ, $-\infty \leq \lambda \leq +\infty$, denotes the spectral function of G. Then $(\mathcal{K}_+,[\cdot,\cdot])$ and $(\mathcal{K}_-,-[\cdot,\cdot])$ are Hilbert spaces and we have

$$\mathcal{K} = \mathcal{K}_+ \dot{+} \mathcal{K}_- , \; [\mathcal{K}_+,\mathcal{K}_-] = (G(I-E_o)\mathcal{K}, E_o\mathcal{K}) = \{0\}.$$

The corresponding positive definite s.p. (1.2) is given by

$$(x,y) = (|G|x,y)_o \qquad (x,y \in \mathcal{K}).$$

If, in particular, the positive or the negative spectrum of G consists only of a finite number of eigenvalues of finite total multiplicity κ , then $(K, [\cdot,\cdot])$ is a π_κ-space. This holds, e.g., if $G = I + \Gamma$ with some compact selfadjoint operator Γ .

(b) Let $(K_o, (\cdot,\cdot)_o)$ be a Hilbert space and G a bounded selfadjoint, indefinite operator in $(K_o, (\cdot,\cdot)_o)$ with $0 \in \sigma(G)$. We consider the factor space $K_1 := K / N(G)$ where $N(G)$ denotes the kernel of G . On K_1 we define the s.p.-s

$$[x,y] := (Gx,y)_o , \quad (x,y) := (|G|\, x,y)_o \qquad (x,y \in K_1).$$

Denote the completion of K_1 with respect to the norm $(x,x)^{1/2}$ by K . The relation

$$|[x,y]| = |(Gx,y)_o| \leq (|G|\, x,x)^{1/2} \cdot (|G|\, y,y)^{1/2} \qquad (x,y \in K_1)$$

implies that the s.p. $[\cdot,\cdot]$ can be extended by continuity to the whole of K. Then $(K, [\cdot,\cdot])$ is a Krein space.

(c) Let L be a complex linear space, $[\cdot,\cdot]$ a Hermitian sesquilinear form on L . The isotropic part of L is denoted by L_o:

$$L_o := \{ x \in L : [x, L] = \{0\}\} .$$

We suppose that the form $[\cdot,\cdot]$ has a finite number κ of negative squares, which means the following:

Each negative linear manifold in L is of dimension $\leq \kappa$ and at least one such manifold is of dimension κ .

Then the factor space L/L_o is a dense subset of a π_κ-space $(K, [\cdot,\cdot])$ with κ negative squares.

In order to prove this we have to choose a κ-dimensional negative subspace L_- of $L_1 := L/L_o$. Then on $L_-^{\perp} := \{x \in L_1 : [x, L_-] = \{0\}\}$ the s.p. $[\cdot,\cdot]$ is positive. Completing the factor space $L_-^{\perp}$ with respect to the corresponding norm the statement easily follows. For details we refer the reader to [12] . The characterizations of convergent and Cauchy sequences in L_1 are also given there in terms of the original s.p. $[\cdot,\cdot]$, without appealing to a special Hilbert norm on the π_κ-space.

I.3. Some classes of linear operators. Let $\mathcal{K} = (\mathcal{K}, [\cdot,\cdot])$ be a Krein space and S a closed, densely defined linear operator in $\mathcal{K}$. The adjoint S^+ of S in the Krein space $\mathcal{K}$ is defined as follows: $\mathcal{D}(S^+)$ is the set of all $x \in \mathcal{K}$ for which there exists an $x' \in \mathcal{K}$ with the property

$$[Sy,x] = [y,x'] \quad \text{for all} \quad y \in \mathcal{D}(S) , \tag{3.1}$$

and for these x we put $S^+x := x'$. With the Hilbert s.p. (1.2) and the corresponding operator J we have

$$S^+ = JS^*J , \tag{3.2}$$

which follows immediatly from (1.3) and (3.1). Hence S^+ is closed and densely defined.

A closed, densely defined linear operator A in the Krein space $\mathcal{K}$ is called Hermitian if $A \subset A^+$, and selfadjoint if $A = A^+$, the linear operator U in $\mathcal{K}$ is called unitary if $UU^+=U^+U= I$, that is, U maps $\mathcal{K}$ onto itself preserving the s.p.$[\cdot,\cdot]$:$[Ux,Uy] =[x,y]$ $(x,y \in \mathcal{K})$. Then U is necessarily continuous [1].

As in Hilbert spaces, selfadjoint and unitary operators in the Krein space $\mathcal{K}$ are connected by the Cayley transformation. As a rule, each statement about selfadjoint operators with nonempty resolvent set in a Krein space has an analogue for unitary operators ; for details see [1] and [12].

The relation (3.2) implies that adjoints in a Krein space satisfy similar relations as adjoints in a Hilbert space, e.g.

$$(\alpha S)^+ = \bar{\alpha} S^+, (S_1+S_2)^+ \supset S_1^+ + S_2^+, (S_1S_2)^+ \supset S_2^+ S_1^+ .$$

Here $\alpha \in \mathbb{C}$, $S, S_1, S_2, S_1 + S_2$ and S_1S_2 are supposed to be closed and densely defined operators in $\mathcal{K}$. In particular, if $\lambda \in \rho(S)$[1]) we have

$$((S - \lambda I)^{-1})^+ = (S^+ - \bar{\lambda} I)^{-1} , \tag{3.3}$$

which yields $\sigma(S^+) = \sigma(S)^*$.[2])

---

[1]) The resolvent set $\rho(S)$ and the spectrum $\sigma(S)$ as well as its subsets $\sigma_p(S)$, $\sigma_c(S)$, $\sigma_r(S)$ are defined as in [14], [1].

[2]) If $\Lambda \subset \mathbb{C}$ we denote $\Lambda^* := \{\bar{\lambda} : \lambda \in \Lambda\}$.

PROPOSITION 3.1. If E is a bounded linear projector in K ($E^2=E$), then the spaces EK and E^+K are in duality with respect to the s.p. $[\cdot, \cdot]$.

That is, for each $x \in EK$, $x \neq 0$, there exists an $x' \in E^+K$ and for each $y \in E^+K$ there exists an $y' \in EK$ such that

$$[x,x'] = [y,y'] = 1 .$$

Indeed, assume e.g. $[x_o,x'] = 0$ for some $x_o \in EK$ and all $x' \in E^+K$. Then we have

$$\{0\} = [x_o,E^+K] = [Ex_o, K], \quad \text{hence} \quad Ex_o = x_o = 0 .$$

Let S be again a closed, densely defined linear operator in K and σ a bounded spectral set of S. The corresponding Riesz projector $E(S;\sigma)$ is given by the relation

$$E(S;\sigma) := -\frac{1}{2\pi i} \int_{C_\sigma} (S-zI)^{-1} dz$$

where C_σ is a closed oriented simple Jordan contour in $\rho(S)$ such that σ is inside and $\sigma(S) \setminus \sigma$ is outside C_σ. Then we have

$$E(S,\sigma)^+ = \left(-\frac{1}{2\pi i} \int_{C_\sigma} (S-zI)^{-1} dz\right)^+ = -\frac{1}{2\pi i} \int_{C_{\sigma^*}} (S^+-zI)^{-1}dz = E(S^+;\sigma^*).$$

The operator $E^+J|_{EK}$ maps EK continuously and bijectively onto E^+K. Indeed, if $\{0\} = (E^+JEK, y_o)$ for some $y_o \in E^+K$ we obtain $E^*EJy_o = 0$, hence $EJy_o = 0$. Observing that $y_o = E^+y_o$ it follows

$$0 = EJE^+y_o = EE^*Jy_o$$

and, finally, $Jy_o = JE^+y_o = E^*Jy_o = 0$, hence $y_o = 0$. Thus the range E^+JEK is dense in E^+K. Let (x_n) be a sequence in EK such that $\| x_n \| = 1$ and $\| E^+Jx_n \| \to 0$ $(n \to \infty)$. Then we have $\| E^*Ex_n \| = \| E^*x_n \| = \| JE^*x_n \| = \| E^+Jx_n \| \to 0$, and $Ex_n = x_n \to 0$ $(n \to \infty)$ follows, a contradiction.

If (e_n), $n = 1,2,\ldots,n_o \leq \infty$ is an orthonormal basis (with respect to the s.p. $(\cdot,\cdot)$) in EK, the vectors $e_n^+ := E^+Je_n$ form a basis in E^+K. Moreover, we have $[e_n, e_m^+] = \delta_{nm}$. With respect to the basis (e_n) the operator $S|_{EK}$ has the matrix representation $S = (\sigma_{jk})$, $\sigma_{jk} = [Se_k, e_j^+]$. If in E^+K the basis (e_n^+) is chosen, then the corresponding matrix of $S^+|_{E^+K}$ is the Hermitian adjoint S^* of S.

If λ_o is an isolated point of $\sigma(S)$ and

$$(S - \lambda I)^{-1} = \sum_{-\infty}^{\infty} (\lambda_o - \lambda)^{\nu} S_{\nu}$$

is the corresponding Laurent expansion of the resolvent of S at λ_o, the relation (3.3) implies

$$(S^{+} - \lambda I)^{-1} = \sum_{-\infty}^{\infty} (\bar{\lambda}_o - \lambda)^{\nu} S_{\nu}^{+} \quad .$$

In particular, the Jordan structures of the corresponding eigenspaces are the same.

An immediate consequence of these considerations is the following

PROPOSITION 3.2. <u>The spectrum</u> $\sigma(A)$ <u>of the selfadjoint operator</u> A <u>in the Krein space</u> K <u>is symmetric with respect to the real axis</u>: $\sigma(A) = \sigma(A)^*$. <u>If</u> σ <u>is a spectral set of</u> A <u>such that</u> $\sigma = \sigma^*$ $(\sigma \cap \sigma^* = \emptyset)$ <u>then</u>

$$E(\sigma;A) = E(\sigma;A)^{+} \qquad (E(\sigma;A)^{+}E(\sigma;A) = 0, \tag{3.4}$$

<u>respectively</u>).

<u>The Jordan structures of the eigenspaces corresponding to isolated eigenvalues</u> $\lambda_o, \bar{\lambda}_o \in \sigma(A)$ <u>are the same</u> .

The second relation in (3.4) implies $E(\sigma;A)K \subset P_o$. For a more detailed study of the spectrum of selfadjoint operators in Krein spaces and, in particular, in Pontrjagin spaces we refer the reader to [1], [12]. Here we only mention that besides the symmetry condition $\sigma(A) = \sigma(A)^*$ (which, in fact, holds also for $\sigma_c(A)$ and $\sigma_p(A)$ separately) the spectrum of a selfadjoint operator in a Krein space can be fairly arbitrary. Indeed, let B be an arbitrary closed operator in a Hilbert space H. In the Hilbert space $(K,(\cdot,\cdot)_o) := H \oplus H$ we define the s.p.

$$[x,y] := \left(\begin{pmatrix} 0 & I \\ I & 0 \end{pmatrix} \begin{pmatrix} x_1 \\ x_2 \end{pmatrix}, \begin{pmatrix} x_1 \\ y_2 \end{pmatrix} \right)_o, \quad x = \begin{pmatrix} x_1 \\ x_2 \end{pmatrix}, \quad y = \begin{pmatrix} y_1 \\ y_2 \end{pmatrix} \quad ,$$

$x_j, y_j \in H$, $j = 1,2$. Then $(K,[\cdot,\cdot])$ is a Krein space, and it is easy to see that the operator

$$A = \begin{pmatrix} B & 0 \\ 0 & B^* \end{pmatrix}$$

is selfadjoint in $(K,[\cdot,\cdot])$ and $\sigma(A) = \sigma(B) \cup \sigma(B^*)$.

A selfadjoint operator A in the Krein space $(K,[\cdot,\cdot])$ is called <u>definitizable</u> if

(1) $\rho(A) \neq \emptyset$.

(2) There exists a real polynomial p such that

$$[p(A)x,x] \geq 0 \qquad (x \in \mathcal{D}(A^k)) \tag{3.5}$$

where $k := \deg p$.

We mention that the reality of the polynomial p does not mean any restriction. The relation (3.5) is evidently equivalent to the following :

$$[p(A)(A-z_oI)^{-k}x,(A-z_oI)^{-k}x] \geq 0 \qquad (x \in K)$$

with $z_o \in \rho(A)$. Moreover, it is also equivalent to the condition (see [5])

(2') There exists a function f which is locally holomorphic on the extended spectrum $\sigma_e(A)$ and does not vanish identically on any spectral set of A and such that

$$[f(A)\,x,x] \geq 0 \qquad (x \in K) .$$

Definitizable operators will be studied in chapter II. Here are some classes of definitizable operators.

(a) A selfadjoint operator in the Krein space K with $\rho(A) \neq \emptyset$, which is <u>nonnegative</u>, that is

$$[Ax,x] \geq 0 \qquad \text{if} \quad x \in \mathcal{D}(A) ,$$

is definitizable.

(b) An arbitrary selfadjoint operator in a π_κ-space is definitizable.

(c) The selfadjoint operator A in the Krein space K with $\rho(A) \neq \emptyset$ is definitizable, if the Hermitian sesquilinear form $[Ax,y]$ $(x,y \in \mathcal{D}(A))$ has a finite number of negative squares.

(d) Let A be a definitizable and B a selfadjoint operator in K. If $\rho(A) \cap \rho(B) \neq \emptyset$ and

$$(A - zI)^{-1} - (B - zI)^{-1}$$

is finite-dimensional for some (and hence for all) $z \in \rho(A) \cap \rho(B)$, then B is definitizable.

Indeed, (a) is obvious. In order to prove (b) we observe that $\rho(A) \neq \emptyset$. If, e.g., $\dim K_- = \kappa$, according to a theorem of

L.S.Pontrjagin there exists a κ-dimensional nonpositive subspace $\mathcal{L}_- \subset \mathcal{D}(A)$ such that $A\mathcal{L}_- \subset \mathcal{L}_-$ ([1], [12]). Let p_o be the minimal polynomial of the restriction $A|_{\mathcal{L}_-}$. Then $p_o(A)\mathcal{L}_- = \{0\}$. This implies (with $k_o := \deg p_o$)

$$[\overline{p}_o(A)\mathcal{D}(A^{k_o}), \mathcal{L}_-] = [\mathcal{D}(A^{k_o}), p_o(A)\mathcal{L}_-] = \{0\},$$

hence

$$\overline{p}_o(A)\mathcal{D}(A^{k_o}) \subset \mathcal{L}_-^{\perp} \subset \mathcal{P}_+ .$$

Here the last inclusion is a consequence of the fact that $\mathcal{L}_-$ is a maximal nonpositive subspace, see [1], [12]. Thus we get $[\overline{p}_o(A)x, \overline{p}_o(A)x] \geq 0$ $(x \in \mathcal{D}(A^{k_o}))$ and (3.5) follows with $p := p_o\overline{p}_o$.

In order to prove (c) we introduce the linear space $\mathcal{L} := \mathcal{D}(A)$ with the Hermitian sesquilinear form $\{x,y\} := [Ax,y]$ $(x,y \in \mathcal{L})$. Let $\mathcal{L}_o$ be its isotropic subspace, $\mathcal{L}_o = \{x : [x,\mathcal{L}] = \{0\}\}$. The factor space $\mathcal{L}/\mathcal{L}_o$ is a dense subset of a π_κ-space $\hat{\mathcal{K}}$, see section 2, (c). Moreover, the set $\mathcal{D}(A^2)$ is dense in $\hat{\mathcal{K}}$. Indeed, let $z \neq \overline{z}$, $z \in \rho(A)$. Then for arbitrary $x \in \mathcal{D}(A)$ there exists a sequence $(x_n) \subset \mathcal{D}(A^2)$ such that $x_n \to x$, $Ax_n \to Ax$ in $\mathcal{K}$ (If $x = (A-zI)^{-1}x' \in \mathcal{K}$, we can choose a sequence $(x_n') \subset \mathcal{D}(A)$, such that $x_n' \to x'$, and set $x_n := (A-zI)^{-1}x_n'$). This implies

$$\{x_n,x_n\} \to \{x,x\}, \quad \{x_n,u\} \to \{x,u\} \qquad (u \in \mathcal{L}),$$

hence $x_n \to x$ in $\hat{\mathcal{K}}$, see [12]. On $\mathcal{D}(A^2)$ we define the operator $\hat{A}$ in $\hat{\mathcal{K}}$: $\hat{A}x = Ax$ $(x \in \mathcal{D}(A^2))$. Then $\hat{A}$ is Hermitian in $\hat{\mathcal{K}}$, and the range $(\hat{A}-zI)\mathcal{D}(A^2) = \mathcal{D}(A)$ is dense in $\hat{\mathcal{K}}$. Therefore $\hat{A}$ is even selfadjoint in $\hat{\mathcal{K}}$ and, according to (b), there exists a polynomial p such that $\{p(\hat{A})x,x\} \geq 0$ $(x \in \mathcal{D}(A^{k+1}))$. This implies $[Ap(A)x,x] \geq 0$ $(x \in \mathcal{D}(A^{k+1}))$.

For the proof of (d) we refer to [15].

I.4. <u>EXAMPLES</u>. (a) A large class of symmetric differential equations can be reduced to a canonical form, which e.g. in the homogeneous case looks as follows:

$$\hat{J}\dot{x}(t) = H(t)x(t), \tag{4.1}$$

see [16],[17]. Here x is an n-vector function, H is a locally summable $n \times n$-matrix function with $H(t) = H(t)^*$ and $\hat{J}$ is a constant $n \times n$ -matrix with the properties

$$\hat{J}^2 = -I \ , \quad \hat{J}^* = -\hat{J} \ .$$

In particular, the canonical equations of classical mechanics on the 2k-dimensional space of generalized coordinates $q_1,\dots,q_k$ and impulses $p_1,\dots,p_k$ with energy $H(p_1,\dots,p_k,q_1,\dots,q_k)$ being a quadratic form in the p_j,q_j, $j = 1,2,\dots,k$ are of the form (4.1) with

$$\hat{J} := \begin{pmatrix} 0 & -I_k \\ I_k & 0 \end{pmatrix} .$$

Consider, e.g. , (4.1) on an interval $[0,L)$, $0 < L \leq \infty$. The solution $U(t)$ of the $n \times n$-matrix differential initial problem

$$\hat{J}\,\dot{U}\,(t) = H(t)\,U(t) \ , \qquad U(0) = I_n \qquad (4.2)$$

is called the <u>matrizant</u> of (4.1). With $J := i\hat{J}$, this matrizant satisfies the relation

$$U(t)^*\,J\,U(t) = J \ .$$

Indeed, we have

$$\frac{d}{dt}\,(U(t)^*\,J\,U(t)) = \dot{U}(t)^*\,i\hat{J}\,U(t) + U(t)^*\,i\hat{J}\,\dot{U}(t) =$$

$$= -i\,U(t)^*\,H(t)\,U(t) + iU(t)^*\,H(t)\,U(t) = 0$$

Introducing the finite-dimensional Krein space $(C^n,[\cdot,\cdot])$ with $[x,y] := i\,(\hat{J}x,y)_n$ $(x,y \in C^n$, $(\cdot,\cdot)_n$ denotes the usual s.p. in $C^n)$ it follows

$$U(t)^+\,U(t) = U(t)\,U(t)^+ = I_n \ .$$

Hence $U(t)$ is unitary in this Krein space.

We mention that other classes of operators also arise in this way. E.g., if we replace the differential equation in (4.2) by

$$J\,\dot{U}(t) = z\,H(t)\,U(t)$$

with some complex parameter z and denote the solution of the corresponding initial problem by $U(t;z)$ $(U(0;z) = I_n)$ it follows

$$U(t;z)^*\,J\,U(t;z) \leq J \quad \text{if} \quad \operatorname{Im} z \geq 0 \ ,$$

which can be written as

$$[U(t;z)x\ ,\ U(t;z)x] \leq [x,x] \qquad (x \in C^n)\ .$$

That is, $U(t;z)$ is a "contractive" operator in the Krein space $(C^n,[\cdot\,,\cdot])$.

Now suppose that the matrix function H in (4.1) is periodic: $H(t+T) = H(t)$ for some $T > 0$ and all $t \geq 0$. Then the stability behavior of the solutions of (4.1) can be studied by means of the monodromy matrix $U(T)$. E.g., the system (4.1) is stable if and only if $\| U(T)^n \| \leq k$ for all $n = 1,2,\ldots$ and some $k > 0$, and strongly stable if this holds for all matrices in a "neighbourhood" of $U(T)$ which are unitary in the Krein space $(C^n, [\cdot\,,\cdot])$. The s.p. $[\cdot\,,\cdot]$ on C^n gives the possibility to classify the eigenvectors and eigenvalues of $U(T)$: Eigenvectors can be positive, negative etc. In these terms necessary and sufficient conditions for the stability of the system (4.1) can be formulated, see [10], [11] and the literature quoted there.

(b) Consider the integral operator

$$(Ax)(t) := \int_0^1 K(t,s)\ x(s)\ d\sigma(S)$$

with some function σ of bounded variation on $[0,1]$ and a continuous Hermitian kernel K on $[0,1]$:

$$K(s,t) = \overline{K(t,s)} \qquad (0 \leq s,\ t \leq 1\ )\ .$$

(see [19]). By L^2_σ we denote the Krein space, consisting of all functions x on $[0,1]$ such that $\int_0^1 |x(t)|^2\ d|\sigma|(t) < \infty$ ($|\sigma|(t)$ denotes the total variation of σ on $[0,t]$) and with s.p. $[x,y] := \int_0^1 x(t)\ \overline{y(t)}\ d\sigma(t)$. Then A is a bounded (even compact) selfadjoint operator in L^2_σ . If the kernel K is positive definite, that is, for arbitrary $n;\ t_1,\ldots,t_n \in [0,1]$ the form

$$\sum_{j,k=1}^{n} K(t_j,t_k)\ \xi_j\ \overline{\xi}_k \tag{4.3}$$

is nonnegative, then the operator A is nonnegative in the Krein space L^2_σ . If the kernel K has a finite number κ of negative squares, that is each of the forms (4.3) has at most κ and at least one of these forms has exactly κ negative squares, then also the Hermitian sesquilinear form $[Ax,y]$ $(x,y \in L^2_\sigma)$ has κ negative squares. According to section 3, (c), the operator A is definitizable.

(c) Operator polynomials (or pencils) in Hilbert space are expressions of the form

$$L(\lambda) = \lambda^n B_n + \lambda^{n-1} B_{n-1} + \dots + \lambda B_1 + B_o , \qquad (4.4)$$

where $\lambda \in C$ is a complex parameter and B_j, $j = 0,1,\dots,n$, are bounded operators in some Hilbert space K. E.g., the study of small damped oscillations often leads to an expression of the form (4.4) with $n = 2$ and selfadjoint operators B_j [20]. Other examples arise in the study of wave guides, see the recent papers [21], [22].

The spectrum, eigenvalues etc. of L are defined as follows: $\lambda \in C$ belongs to the spectrum $\sigma(L)$ ($\sigma_p(L)$ etc.) if $0 \in \sigma(L(\lambda))$ ($\in \sigma_p(L(\lambda))$ etc.). Assume now $B_n = I$. It is easy to check that the spectrum etc. of L coincides with the spectrum of the operator

$$A: = \begin{pmatrix} -B_{n-1} & -B_{n-2} & \dots & -B_1 & -B_o \\ I & 0 & \dots & 0 & 0 \\ 0 & I & \dots & 0 & 0 \\ 0 & 0 & \dots & I & 0 \end{pmatrix}$$

in the space $K = H^n$ 1). If, moreover, the operators $B_o, B_1, \dots, B_{n-1}$ are even selfadjoint in H, it is easy to check that the operator A is selfadjoint in the Krein space $(K, [\cdot,\cdot])$ where

$$[x,y]: = (Gx,y) \qquad (x,y \in K),$$

and

$$G: = \begin{pmatrix} 0 & 0 & \dots & 0 & I \\ 0 & 0 & \dots & I & B_{n-1} \\ \vdots & \vdots & & \vdots & \vdots \\ 0 & I & \dots & B_3 & B_2 \\ I & B_{n-1} & \dots & B_2 & B_1 \end{pmatrix}$$

In case $n = 2$ the operator polynomial (4.4) with $B_2 = I$ and B_o, B_1 selfadjoint in H is said to be <u>strongly damped</u> if

$$(B_1 x,x)^2 \geq 4 (B_o x,x) \| x \|^2 \qquad (x \in H),$$

---

1) If $H = C$, that is $B_o, B_1, \dots, B_{n-1}$ are complex numbers, $B_n = 1$, it is well-known that $L(\lambda)$ is the characteristic polynomial of the matrix A.

that is, each polynomial $\lambda \to (L(\lambda)x,x)$ $(x \in H,\ x \neq 0)$ has only real zeros. It turns out that in this case there exists a real λ_o such that

$$[(A - \lambda_o I)x,x] \geq 0 \qquad (x \in K)$$

holds. More generally, if $n \geq 2$, $B_n = I$ and each polynomial $\lambda \to (L(\lambda)x,x)$ has n real zeros then A is definitizable with a definitizing polynomial of degree $\leq n-1$, see [23] and other papers quoted in [11].

(d) Consider the Sturm-Liouville differential equation

$$-(p^{-1}y')' + q\,y = \lambda\,r\,y \quad \text{on } [0,\infty)^{1)} \qquad (4.5)$$

in the so-called "polar case"; that means we have $p \geq 0$, $q > 0$, but r can change its sign. It was studied by H.Weyl [24] with initial conditions $y(0) = 0$ or $y'(0) = 0$. Here we impose the more general initial condition

$$y(0)\cos\alpha - p(0)\,y'(0)\sin\alpha = 0 \quad \text{with some } 0 \leq \alpha < \pi. \qquad (4.6)$$

Multiplying the expression on the left hand side in (4.5) by $\bar{y}$ and integrating along the nonnegative axis we get for a function y which vanishes at ∞

$$\int_0^\infty (p^{-1}|y'|^2 + q\,y^2)\,dt + \cot\alpha\,|y(0)|^2 \ ; \qquad (4.7)$$

here the last term is zero if $\alpha = 0$ or $\alpha = \pi/2$. Doing the same with the factor of λ on the right hand side of (4.5) we get

$$\int_0^\infty r\,|y|^2\,dt \ .$$

If r changes its sign, the s.p. $[x,y] := \int_0^\infty r\,x\,\bar{y}\,dt$ need not be positive definite. However, in polar case, if $0 \leq \alpha \leq \pi/2$, the quadratic form (4.7) is positive definite and can be used in order to define a Hilbert space and, subsequently, a Hermitian or self-adjoint operator in this space can be associated with the problem

---

1) The coefficients p,q,r are supposed to fulfil some basic requirements such that existence theorems hold and operators can be associated with the problem (4.5).

(4.5), (4.6). If, however, α is sufficiently close to π, the s.p. corresponding to (4.7):

$$\{x,y\} := \int_0^\infty (p^{-1}x'\overline{y}' + q\, x\, \overline{y})\, dt + \cot\alpha\, x(0)\, \overline{y(0)}$$

becomes indefinite, more exactly, it has one negative square. Now a Hermitian or selfadjoint operator A in the corresponding π_1-space can be associated with the problem (4.5),(4.6). Moreover, in this case the Krein space $(L^2(r), [\cdot,\cdot])$ can also be considered ($x \in L^2(r)$ if $\int_0^\infty |x|^2 |r|\, dt < \infty$) and an operator A can be associated with (4.5), (4.6) such that the Hermitian sesquilinear form $[Ax,y]$ $(x,y \in \mathcal{D}(A))$ has one negative square. If, additionally, this operator A is selfadjoint and $\rho(A) \neq \emptyset$, it is definitizable, see [4].

Apparently, some vector systems of the form (4.5) or higher order equations of the form

$$S\, x = \lambda\, T\, x$$

with symmetric differential expressions S, T can be treated in a similar way if one of the Dirichlet forms associated with S or T has only a finite number of negative squares.

(e) In the spectral theory of selfadjoint operators in Hilbert space and its applications, e.g. to moment or Sturm-Liouville problems, analytic complex (or matrix) functions f, mapping the upper half plane C_+ into itself, play a fundamental role. These so-called Nevanlinna functions f can be characterized by the property that for arbitrary n and $z_1,\dots,z_n \in C_+$ the Hermitian form

$$\sum_{j,k=1}^{n} \frac{f(z_j)-\overline{f(z_k)}}{z_j-\overline{z}_k}\, \xi_j\, \overline{\xi}_k \tag{4.8}$$

is nonnegative. In a similar way, meromorphic functions f on C_+ such that the Hermitian forms (4.8) have at most κ and at least one such form has exactly κ negative squares, play an important role in the study of the problems mentioned in d) as well as in some other questions. These functions admit realizations by means of selfadjoint or unitary operators in π_κ-spaces and integral representations generalizing the well-known integral representations of Nevanlinna functions.

Here is an example of such a function f :

$$f(z) = \int_{-\infty}^{\infty} \frac{d\sigma(t)}{t-z} - z + \sum_{j=1}^{n} \sum_{\ell=1}^{\ell_j} \left(\frac{\beta_{j,\ell}}{(z-z_j)^{\ell}} + \frac{\overline{\beta}_{j,\ell}}{(z-\overline{z}_j)^{\ell}}\right); \beta_{j,\ell_j} \neq 0$$

σ is a nondecreasing bounded function on R^1, $z_1,\ldots,z_n$ are mutually different points of C_+. The maximal number κ of negative squares of the forms (4.8) is equal to $\kappa = 1 + \sum_{j=1}^{n} \ell_j$. For the details as well as for other classes of complex analytic functions connected with the theory of operators in π_κ-spaces we refer the reader to [3] and other papers quoted there.

Selfadjoint operators in π_κ - spaces have also applications in the theory of eigenvalue problems for differential operators, containing the eigenvalue parameter in the boundary conditions (see [25] and other papers quoted in [11]) as well as in the study of the transport equation [26].

I.5. Orthogonality. Subspaces. If L is an arbitrary subset of a Krein space $(K, [\cdot,\cdot])$ we put

$$L^{\perp} := \{x \in K : [x, L] = \{0\}\}.$$

Fix a decomposition (1.1) and consider the corresponding positive definite s.p. (1.2) and the operator J. Then $[x, L] = \{0\}$ is equivalent to $(Jx, L) = (x, JL) = \{0\}$. Denoting the orthogonal complement with respect to the s.p. $(\cdot,\cdot)$ by $(\perp)$, it follows

$$L^{\perp} = J L^{(\perp)} = (J L)^{(\perp)} . \tag{5.1}$$

Evidently, $L^{\perp}$ is a subspace of K. The following relations are easy to check (L, M, N are arbitrary subsets of K):

$$L \subset (L^{\perp})^{\perp} ; \quad M \subset L \Rightarrow M^{\perp} \supset L^{\perp} ; \tag{5.2}$$

$$(M + N)^{\perp} = M^{\perp} \cap N^{\perp} ; \quad (M \cap N)^{\perp} \supset M^{\perp} + N^{\perp} .$$

Now suppose that L is even a subspace of $(K, [\cdot,\cdot])$. Then the equality

$$(L^{\perp})^{\perp} = L \tag{5.3}$$

holds. Indeed, (5.1) yields

$$(L^{\perp})^{\perp} = (J L^{(\perp)})^{\perp} = (L^{(\perp)})^{(\perp)} = L .$$

The isotropic part $L_o = L \cap L^{\perp}$ is a subspace of K contained in P_o, and we have from (5.3) and (5.2)

$$L_o = (L + L^{\perp})^{\perp}$$

The following proposition is an easy consequence of this equality.

PROPOSITION 5.1. The relations

$$L \cap L^{\perp} = \{0\} \quad \text{and} \quad \overline{L+L^{\perp}} = K$$

are equivalent. If they hold the sum in the second equality is even a direct one.

The projector E ($E^2 = E$) in $(K, [\cdot,\cdot])$ is called orthogonal if $E = E^+$. This is equivalent to $(EK)^{\perp} = (I - E)K$, and from Proposition 3.1 we know that $EK \cap (EK)^{\perp} = \{0\}$.

THEOREM 5.2. If L is a subspace of the Krein space $(K, [\cdot,\cdot])$, the following assertions are equivalent:

(a) $(L, [\cdot,\cdot])$ is a Krein space.

(b) $(L^{\perp}, [\cdot,\cdot])$ is a Krein space.

(c) $K = L \dot{+} L^{\perp}$.

(d) L is the range of an orthogonal projector E in $K : L = EK$.

PROOF. (a) $\Rightarrow$(c): If $(L, [\cdot,\cdot])$ is a Krein space it follows from (iii) in section 1 that $L \cap L^{\perp} = \{0\}$, and the Proposition 5.1 implies $L \dot{+} L^{\perp} = K$. Therefore, for an arbitrary $x \in K$ there exist sequences $(x_n') \subset L$, $(x_n'') \subset L^{\perp}$, such that $x_n' + x_n'' \to x$ $(n \to \infty)$ in K. This implies $[x_n', y'] \to [x, y']$ $(n \to \infty)$ for all $y' \in L$. In particular, (x_n') is a weak Cauchy sequence in the Hilbert space $(L, (\cdot,\cdot)')$, where $(\cdot,\cdot)'$ is a positive definite s.p. on L defined by a decomposition (1.1) of L: $L = L_+ \dot{+} L_-$, according to (1.2). Since Hilbert space is weakly complete, there exists some $x_o' \in L$ such that $x_n' \to x_o'$ $(n \to \infty)$ weakly with respect to the s.p. $(\cdot,\cdot)'$. Moreover,

$$[x - x'_o, L] = \lim_{n\to\infty} [x'_n + x''_n - x'_o, L] = \{0\},$$

that is, $x - x'_o \in L^\perp$. Thus, $x \in K$ admits the decomposition $x = x'_o + (x - x'_o)$, $x'_o \in L$, $x - x'_o \in L^\perp$, and (c) holds.

(c) ⟹ (a) : We consider the s.p. (1.2), which is defined on K, on the subspace L. Then $(L, (\cdot,\cdot))$ is a Hilbert space and we have

$$|[x,y]| \leq (x,x)^{1/2} (y,y)^{1/2} \qquad (x,y \in L).$$

Hence there exists a bounded selfadjoint operator G_L in L such that $[x,y] = (G_L x,y)$ $(x,y \in L)$. The statement (a) follows if we show that $0 \in \rho(G_L)$ (see section 2, (a)). Evidently, $0 \in \sigma_p(G_L)$ is impossible as the s.p. $[\cdot,\cdot]$ is nondegenerated on L. Assume $0 \in \sigma_c(G_L)$. Then there exist an element $y_o \in L$, $y_o \notin G_L L$, and a sequence $(y_n) \subset G_L L$ such that $y_n \to y_o$ $(n \to \infty)$ with respect to the Hilbert norm on K (or L). Let $y_n = G_L x_n, x_n \in L$, $n = 1,2,\ldots$. It follows that

$$[x_n,y] = (G_L x_n,y) \to (y_o,y) \qquad (n \to \infty,\ y \in L) \tag{5.4}$$

and $[x_n, L^\perp] = \{0\}$. According to (c), (x_n) is a weak Cauchy sequence in K. Hence there exists an $x_o \in K$ such that $x_n \to x_o$ $(n \to \infty)$ weakly in K. The element x_o belongs even to L and

$$[x_n,y] \to [x_o,y] = (G_L x_o,y) \qquad (n \to \infty,\ y \in L).$$

Comparing this relation with (5.4) it follows $y_o = G_L x_o$, a contradiction.

(c) ⟺ (d): If a decomposition (c) is given, $x = x' + x''$ with $x' \in L$, $x'' \in L^\perp$ for arbitrary $x \in K$, we define

$$Ex := x'.$$

Then, as L and $L^\perp$ are subspaces, E is bounded, $E^2 = E$ and $E^+ = E$. The converse (d) ⟹ (c) is easy to see.

Finally, if we replace in (c) L by $L^\perp$ the equivalence with (b) follows.

REMARK 1. If L is even a nonnegative subspace of $(K, [\cdot,\cdot])$ then (a) can be replaced by

(a_+) $(\mathcal{L}, [\cdot, \cdot])$ is a Hilbert space

and the assertions (a) - (d) are also equivalent to

(e_+) $\mathcal{L}$ is a uniformly positive subspace of $\mathcal{K}$.

The simple proof of these facts can be left to the reader.

REMARK 2. A linear manifold $\mathcal{L}$ in $\mathcal{K}$ for which (c) or (d) hold is automatically closed, that is, a subspace. If, however, we suppose (a) or (b) then $\mathcal{L}$ need not be closed in $\mathcal{K}$. Examples, even in a π_1-space $\mathcal{K}$ and with a definite manifold $\mathcal{L}$, can be found in [1].

Later we shall need the following simple lemma.

LEMMA 5.3. <u>If</u> E_1, E_2 <u>are two commuting orthogonal projectors in the Krein space</u> $\mathcal{K}$ <u>and</u> $E_j \mathcal{K} \subset \mathcal{P}_+$, $j = 1,2$, <u>then also</u> $E_1\mathcal{K} + E_2\mathcal{K} \subset \mathcal{P}_+$.

PROOF. The ranges of the orthogonal and mutually orthogonal projectors E_1, $E_2 - E_1E_2$ are contained in $\mathcal{P}_+$. Hence, the linear span of these ranges is also in $\mathcal{P}_+$. On the other hand this linear span equals $E_1\mathcal{K} + E_2\mathcal{K}$.

II. SPECTRAL FUNCTIONS OF DEFINITIZABLE OPERATORS

II.1. Inversion formulas. Let $(\mathcal{H}, (\cdot,\cdot))$ be a Hilbert space and F a function on R^1 whose values are bounded linear operators in $\mathcal{H}$ with the following properties:

(1) $F(t) = F(t)^*$ $\quad (t \in R^1)$;

(2) $s \leq t$ implies $F(s) \leq F(t)$ $\quad (s,t \in R^1)$;

(3) F is bounded : $\|F(t)\| \leq k$ $\quad (t \in R^1)$.

As a conseqeunce of (1)-(3) the limits $F(t \pm 0)$ $(t \in R^1)$, $F(\pm\infty)$, exist in the strong operator topology. We denote

$$\tilde{F}(t) := \frac{F(t+0)+F(t-0)}{2} \qquad (t \in R^1)$$

and suppose in the following that $F(-\infty) = 0$.

The integral

$$\Gamma(z) := \int_{-\infty}^{\infty} (t-z)^{-1}\, dF(t) \qquad (z \neq \bar{z}) ,$$

(considered in the weak, strong or uniform operator topology) defines an operator function Γ which is holomorphic on the open upper and lower half planes as well as on those intervals of the real axis where F is constant.

If $\Delta = [\alpha_1, \alpha_2]$ is a finite closed interval of the real axis, by C_Δ we denote the closed oriented curve in the complex plane which consists of the line segments connecting the points $\alpha_2 + i\beta$, $\alpha_1 + i\beta$, $\alpha_1 - i\beta$, $\alpha_2 - i\beta$, $\alpha_2 + i\beta$ for some $\beta > 0$. If we take off from C_Δ the segments between $\alpha_1 + i\delta$, $\alpha_1 - i\delta$ and $\alpha_2 - i\delta$, $\alpha_2 + i\delta$ $(0 < \delta < \beta)$ the remaining part of C_Δ will be denoted by C_Δ^δ.

LEMMA 1.1. For each $x \in \mathcal{H}$ we have

$$\lim_{\delta \downarrow 0} \left(-\frac{1}{2\pi i} \int_{C_\Delta^\delta} \Gamma(z)\, x\, dz\right) = \int_\Delta d\tilde{F}(t)\, x ,$$

where the limit exists in the norm of $\mathcal{H}$.

PROOF[1]. We have

$$-\frac{1}{2\pi i}\int_{C_\Delta^\delta}\Gamma(z)x\,dz = $$
$$=\int_{-\infty}^{\infty}\Big(-\frac{1}{2\pi i}\int_{C_\Delta^\delta}(t-z)^{-1}\,dz\Big)\,d\,F(t)x$$

and

$$-\frac{1}{2\pi i}\int_{C_\Delta^\delta}(t-z)^{-1}\,dz =$$
$$=-\frac{1}{2\pi i}\Big(\int_{\alpha_2}^{\alpha_1}(t-s-i\delta)^{-1}\,ds+\int_{\alpha_1}^{\alpha_2}(t-s+i\delta)^{-1}ds\Big) =$$
$$=\frac{\delta}{\pi}\int_{\alpha_1}^{\alpha_2}\frac{ds}{(t-s)^2+\delta^2}=\frac{1}{\pi}\ \mathrm{arc\ tan}\ \frac{\alpha_2-t}{\delta} -$$
$$-\frac{1}{\pi}\ \mathrm{arc\ tan}\ \frac{\alpha_1-t}{\delta}\ .$$

Furthermore,

$$\frac{1}{\pi}\int_{-\infty}^{\infty}\mathrm{arc\ tan}\ \frac{\alpha_2-t}{\delta}\ dF(t)\ x =$$
$$=\frac{1}{\pi}\int_{-\infty}^{\alpha_2-0}\mathrm{arc\ tan}\ \frac{\alpha_2-t}{\delta}\ dF(t)x+\frac{1}{\pi}\int_{\alpha_2+0}^{\infty}\mathrm{arc\ tan}\ \frac{\alpha_2-t}{\delta}dF(t)\ x\ .$$

If $\delta\downarrow 0$ this tends to

$$\frac{1}{2}\int_{-\infty}^{\alpha_2-0}dF(t)x-\frac{1}{2}\int_{\alpha_2+0}^{\infty}dF(t)x$$

in the norm of H. In the same way we get

$$\lim_{\delta\downarrow 0}\Big(-\frac{1}{\pi}\int_{-\infty}^{\infty}\mathrm{arc\ tan}\ \frac{\alpha_1-t}{\delta}\ dF(t)x\Big) =$$
$$=-\frac{1}{2}\int_{-\infty}^{\alpha_1-0}dF(t)x+\frac{1}{2}\int_{\alpha_1+0}^{\infty}dF(t)x\ ,$$

---

1) I thank P.Jonas for communicating this proof to me.

and the statement follows easily.

COROLLARY 1. *If* $\Delta_\varepsilon := [\alpha_1 \mp \varepsilon, \alpha_2 \pm \varepsilon]$, *for arbitrary* $x \in H$ *the limit*

$$\lim_{\varepsilon \downarrow 0} \lim_{\delta \downarrow 0} \left(-\frac{1}{2\pi i} \int_{C^\delta_{\Delta_\varepsilon}} \Gamma(z)\, x\, dz\right)$$

exists in the norm of H *and equals* $\int_{\alpha_1 \mp 0}^{\alpha_2 \pm 0} dF(t)\, x$.

COROLLARY 2. *Let* f *be a complex function which is holomorphic on* C_Δ. *Then for arbitrary* $x \in H$ *the limits*

$$\lim_{\delta \downarrow 0} \left(-\frac{1}{2\pi i} \int_{C^\delta_\Delta} f(z)\, \Gamma(z) x\, dz\right) ,$$

$$\lim_{\varepsilon \downarrow 0} \lim_{\delta \downarrow 0} \left(-\frac{1}{2\pi i} \int_{C^\delta_{\Delta_\varepsilon}} f(z)\, \Gamma(z) x\, dz\right)$$

exist in the norm of H . *If* f *is even holomorphic on* $\bar{\Delta}$ *they are equal to*

$$\int_{\alpha_1}^{\alpha_2} f(t)\, d\tilde{F}(t) \quad \textit{and} \quad \int_{\alpha_1 \mp 0}^{\alpha_2 \pm 0} f(t)\, dF(t), \textit{ respectively.} \qquad (1.1)$$

In order to see this we choose closed intervals Δ'_1 , Δ'_2 around α_1 , α_2 such that f is holomorphic on each of these intervals. Consider the relation

$$f(z)\, \Gamma(z) x = f(z) \int_{R^1 \setminus (\Delta'_1 \cup \Delta'_2)} (t-z)^{-1}\, dF(t)\, x +$$

$$+ \sum_{j=1}^{2} \int_{\Delta'_j} (t-z)^{-1} (f(z)-f(t))\, dF(t)\, x + \sum_{j=1}^{2} \int_{\Delta'_j} (t-z)^{-1} f(t)\, dF(t)\, x.$$

As the first two terms on the right hand side are holomorphic on a neighbourhood of C_Δ it remains to apply Lemma 1.1 to the third term. For the proof of (1.1) we can suppose that f is real on the real points of its domain (writing $f = f_r + i f_i$ with f_r and f_i

having this property). If $f(\alpha_j) \neq 0$ then we choose Δ_j' such that sign $f(t) = \text{sign } f(\alpha_j)$ for $t \in \Delta_j'$, and replace in Lemma 1.1 $dF(t)$ by $|f(t)|\, dF(t)$. If $f(\alpha_j) = 0$ we split the sufficiently small interval Δ_j' into two intervals such that f is of constant sign on each of them.

REMARK. The statements of Lemma 1.1 and its corollaries remain true if F is the difference of two nondecreasing functions or, more generally, if it is an operator function of bounded weak variation.

II.2. <u>A representation of the resolvent</u>. Let A be a definitizable operator in the Krein space $(K, [\cdot,\cdot])$ with definitizing polynomial p:

$$[p(A)x,x] \geq 0 \quad (x \in \mathcal{D}(A^k),\ k := \deg p).$$

We fix some $z_o \in \rho(A)$ and put

$$q(z) := p(z)\,(z-z_o)^{-k}(z-\bar{z}_o)^{-k}.$$

Then the only singularities of q in the closed complex plane are poles at z_o and $\bar{z}_o$, and the same holds for the function

$$Q(z,\zeta) := (z-\zeta)^{-1}(q(z)-q(\zeta)),\ Q(z,z) := q'(z)$$

$$(z,\zeta \neq z_o,\bar{z}_o; z \neq \zeta)$$

with respect to both z and ζ.

By R_z we denote the resolvent of A: $R_z := (A-zI)^{-1}(z \in \rho(A))$. The operator $q(A) = p(A)\,R_{z_o}^k\,R_{\bar{z}_o}^k$ is bounded and nonnegative in K:

$$[q(A)x,x] \geq 0 \qquad (x \in K).$$

If $z \in \rho(A)$, $z \neq z_o, \bar{z}_o$, the operator

$$Q(z,A) = (q(A) - q(z)\,I)\,R_z \tag{2.1}$$

is also bounded and depends holomorphically on z in $\rho(A)$. Moreover, the function $Q(z,A)$ has a holomorphic extension to the closed complex plane with the exception of the points z_o, $\bar{z}_o$. This follows from the Riesz-Dunford-Taylor functional calculus for closed operators or from the identity

$$Q(z,\zeta) = p(\zeta)(\zeta-\bar{z}_o)^{-k}\sum_{j=0}^{k-1}(\zeta-z_o)^{-j}(z-z_o)^{j-k+1}$$
$$+\,(z-z_o)^{-k}(z-\zeta)^{-1}(p(z)-p(\zeta))(\zeta-\bar{z}_o)^{-k}$$
$$+\,(z-z_o)^{-k}\,p(z)\sum_{j=0}^{k-1}(\zeta-\bar{z}_o)^{-j}(z-\bar{z}_o)^{j-k+1}\,.$$

We denote the values of this extension also by $Q(z,A)$ and observe, that (2.1) yields the relation

$$R_z = q(z)^{-1}\,q(A)R_z - q(z)^{-1}\,Q(z,A)\quad (z\in\rho(A), q(z)\neq 0)\,. \qquad (2.2)$$

If $x \in K$ we define the function F_x :

$$F_x(z) := [\,q(A)\ R_z x, x\,] \qquad (z\in\rho(A))\,.$$

Then

$$\operatorname{Im} F_x(z) = \operatorname{Im}[q(A)\ R_z x,x] = \eta[q(A)\ R_z x, R_z x] \quad (\eta = \operatorname{Im} z) \quad (2.3)$$

which, by Schwartz' inequality, implies ·

$$\eta[q(A)R_z x, R_z x\,] \leq |[q(A)R_z x,x]| \leq [q(A)\ R_z x,R_z x\,]^{1/2}\,[q(A)x,x]^{1/2}$$

or

$$\eta^2\,[q(A)R_z x,\ R_z x] \leq [\,q(A)\ x,x]\,.$$

Thus the relation (2.3) yields

$$\operatorname{Im} F_x(z) \leq \frac{1}{\eta}\,[q(A)x,x] \qquad (x\in K)\,.$$

Moreover, we shall show that for $x,y \in K$

$$-i\eta\,[q(A)\ R_{i\eta}\,x,y] \to [\,q(A)x,y\,] \qquad (\eta\uparrow\infty)\,, \qquad (2.4)$$

which means

$$-i\eta\,F\,(i\eta) \to [\,q(A)\ x,x\,] \qquad (\eta\uparrow\infty)\,. \qquad (2.5)$$

In order to prove (2.4) we observe that it holds if $x \in \mathcal{D}(A)$:

$$|[q(A)x,y] + i\eta\,[q(A)R_{i\eta}\,x,y]|^2 = |[q(A)\;A\;R_{i\eta}\,x,y]|^2 \leq$$

$$\leq [q(A)\;Ax,\;Ax]\,[q(A)\;R_{i\eta}\;y,R_{i\eta}\;y] \to 0 \quad (\eta\uparrow\infty)$$

and that for arbitrary $x \in K$ we have

$$|i\eta\;[q(A)\;R_{i\eta}\;x,y]|^2 \leq \eta^2[q(A)\;R_{i\eta}\;x,\;R_{i\eta}\;x]\;[\,q(A)\;y,y\,] \leq$$

$$\leq [q(A)\;x,x]\;[q(A)\;y,y\,]\;.$$

According to [27, §4] the function F_x, mapping the upper half plane into itself and satisfying (2.5), admits an integral representation

$$F_x(z) = \int_{-\infty}^{\infty} (t-z)^{-1}\,d\,\sigma_x(t) \qquad (z\neq\bar{z})$$

with some nondecreasing function σ_x on R^1 such that

$$\sigma_x(-\infty) = 0\;,\;\sigma_x(t) = \sigma_x(t-0)\;\;(t\in R^1),\;\;\sigma_x(+\infty) = [q(A)\;x,x\,].$$

By standard arguments, for arbitrary $x,y \in K$ the function

$F_{x,y}(z) := [q(A)\;R_z x,y\,]$ admits a representation

$$F_{x,y}(z) = \int_{-\infty}^{\infty} (t-z)^{-1}\,d\,\sigma_{x,y}(t) \qquad (z\neq\bar{z}) \qquad (2.6)$$

with some left-continuous function $\sigma_{x,y}$ of bounded variation in R^1, which for fixed t is a Hermitian sesquilinear form with respect to x,y,

$$\sigma_{x,y}(-\infty) = 0\;,\;\;\sigma_{x,y}(\infty) = [\,q(A)\;x,y\,] \qquad (2.7)$$

$$|\,\sigma_{x,y}(t)\,|^2 \leq \sigma_x(t)\;\sigma_y(t) \leq [q(A)\;x,x]\;[\,q(A)\;y,y\,]\,.$$

If we choose a Hilbert s.p. $(\cdot,\cdot)$ on K according to (I.1.1), (I.1.2) it follows that there exists a nondecreasing left-continuous function F on R^1 whose values are bounded selfadjoint operators in the Hilbert space $(K,(\cdot,\cdot))$, such that

$$\sigma_{x,y}(t) = (F(t)\;x,y) \qquad (x,y \in K).$$

Then (2.6) and the second relation in (2.7) imply

$$[q(A)\,R_z\,x,y] = \int_{-\infty}^{\infty} (t-z)^{-1}\,d(F(t)x,y)\ ,$$
$$[q(A)\,x,y] = \int_{-\infty}^{\infty} d(F(t)\,x,y) \qquad (x,y\in K)\ .$$

These equalities can be written in the form

$$q(A)R_z = J\int_{-\infty}^{\infty}(t-z)^{-1}\,d\,F(t)\ ,\ q(A) = J\int_{-\infty}^{\infty} d\,F(t)\ , \qquad (2.8)$$

the integrals exist in the strong operator topology (s.o.t.). Finally, (2.2) yields the representation

$$R_z = q(z)^{-1}\,J\int_{-\infty}^{\infty}(t-z)^{-1}\,d\,F(t) - q(z)^{-1}\,Q(z;A)\ ,(2.9)$$

where the integral also exists in the s.o.t.

Evidently, if $z\in\sigma(A)$ it is either a zero of p or a point of increase of F. Some further simple consequences of (2.9) are formulated in the following proposition. There, for an arbitrary complex number λ we denote by $k(\lambda)$ its multiplicity as a zero of the fixed definitizing polynomial p ($k(\lambda) := 0$ if $p(\lambda)\neq 0$).

PROPOSITION 2.1. Let A be a definitizable operator in the Krein space K with definitizing polynomial p. Then the nonreal spectrum $\sigma_o(A)$ of A consists of a finite number of pairs $\lambda,\bar\lambda$. Each isolated spectral point of A is an eigenvalue of finite Riesz index $\nu(\lambda)=\nu(\bar\lambda)$:

$$\nu(\lambda)\leq\begin{cases}k(\lambda) & \text{if } \lambda\neq\bar\lambda\ ,\\ k(\lambda)+1 & \text{if } \lambda=\bar\lambda\ .\end{cases}$$

Near the real axis the relation

$$\|R_{\xi+i\eta}\| = O(\eta^{-k(\xi)-1}) \qquad (\eta\to 0)$$

holds; in particular, if ξ is an eigenvalue of A each corresponding Jordan chain is of length $\leq k(\xi)+1$.

The following lemma will be needed in order to prove some corollaries of Proposition 2.1.

LEMMA 2.2. If A is a definitizable operator in the Krein space K and $0\notin\sigma_p(A)$ then A^{-1} is also definitizable.

PROOF. The spectral mapping theorem implies that $\rho(A^{-1})\neq\emptyset$.

If $p(\lambda) = \sum_{j=0}^{k} \pi_j \lambda^j$ is a definitizing polynomial of A, it follows with $B := A^{-1}$, $y \in \mathcal{D}(B^{2k})$, $x := B^k y$ ($\in \mathcal{D}(A^k)$) :

$$[\sum_{j=0}^{k} \pi_j B^{2k-j}y, y] = [\sum_{j=0}^{k} \pi_j A^{j-2k}A^k x, A^k x] = [p(A)x, x] \geq 0,$$

and the lemma is proved.

COROLLARY 1. <u>The (finite) spectrum of a definitizable operator A is nonempty.</u>

Indeed, assuming the contrary we consider the definitizable operator A^{-1}. Then 0 is the only point in $\sigma(A^{-1})$, hence an eigenvalue of A^{-1}, which is impossible.

COROLLARY 2. <u>A definitizable operator A with bounded spectrum is bounded.</u>

Indeed, if K_1 is the Riesz spectral subspace of A corresponding to the bounded spectrum $\sigma(A)$, the restriction $A|K_1^{\perp}$ is a definitizable operator with empty spectrum, hence $K_1^{\perp} = \{0\}$.

The Propositions 2.1, I.3.2 and I.5.2 imply that the space K can be decomposed:

$$K = K_0 \dot{+} K_1, \quad K_1 = K_0^{\perp} \tag{2.10}$$

such that $K_0 \subset \mathcal{D}(A)$, $A|K_0$ is bounded and $\sigma(A|K_0)$ coincides with the nonreal spectrum $\sigma_0(A)$ of A, while $\sigma(A|K_1)$ is real. Here K_0 is the Riesz spectral subspace corresponding to the nonreal spectrum $\sigma_0(A)$. Evidently, there exists a polynomial p_0 such that $p_0(A|K_0) = 0$.

II.3. <u>Spectral functions</u>. Let A be a definitizable operator in the Krein space K. We denote

$$c(A) := (\bigcap N(p)) \cap \sigma(A) \cap R^1,$$

where $N(p)$ is the set of zeros of the polynomial p, and the first intersection runs over all definitizing polynomials p of A. The elements of $c(A)$ are called the (finite) <u>critical points</u> of A. In section 4 another characterization of $c(A)$ will be given which does not make use of definitizing polynomials. Further, R_A is the semiring which consists of all bounded intervals of R^1 with endpoints not in $c(A)$ and their complements (in R^1).

THEOREM 3.1. Let A be a definitizable operator in the Krein space K. Then there exists a mapping $\Delta \to E(\Delta)$ from R_A into the set of bounded linear operators in K with the following properties ($\Delta, \Delta' \in R_A$):

(1) $E(\Delta) = E(\Delta)^+$.

(2) $E(\Delta)\, E(\Delta') = E(\Delta \cap \Delta')$.

(3) $E(\Delta \cup \Delta') = E(\Delta) + E(\Delta')$ if $\Delta \cup \Delta' \in R_A$, $\Delta \cap \Delta' = \emptyset$.

(4) $E(\Delta)K$ is a positive (negative) subspace if $p > 0$ ($p < 0$, respectively) on $\bar{\Delta} \cap \sigma(A)$ for some definitizing polynomial p of A.

(5) $E(\Delta)$ is in the double commutant of the resolvent of A.

(6) If Δ is bounded then $E(\Delta)K \subset \mathcal{D}(A)$ and $A|E(\Delta)K$ is a bounded operator.

(7) $\sigma(A|E(\Delta)K) \subset \bar{\Delta}$.

PROOF. Let $\Delta \in R_A$ be closed and bounded. We choose C_Δ (see section 1) such that all the points of the nonreal spectrum $\sigma_o(A)$ are in the exterior of C_Δ and define

$$\tilde{E}(\Delta)x := -\frac{1}{2\pi i}\int_{C_\Delta}' R_z\, x\, dz := \lim_{\delta \downarrow 0}\Big(-\frac{1}{2\pi i}\int_{C_\Delta^\delta} R_z\, dz\Big).$$

Here, according to (2.9) and Corollary 2 of Lemma 1.1, the limit exists in the norm of K.

If $\Delta, \Delta' \in R_A$ are two bounded closed intervals of positive length such that $\Delta \cup \Delta' \in R_A$ and $\Delta \cap \Delta'$ consists only of one common boundary point, then we have evidently

$$\tilde{E}(\Delta \cup \Delta') = \tilde{E}(\Delta) + \tilde{E}(\Delta'). \tag{3.1}$$

We show that for two bounded closed intervals $\Delta, \Delta' \in R_A$ without common endpoints the relation

$$\tilde{E}(\Delta)\, \tilde{E}(\Delta') = \tilde{E}(\Delta \cap \Delta') \tag{3.2}$$

holds. If e.g., Δ and Δ' are located as in the figure, we choose

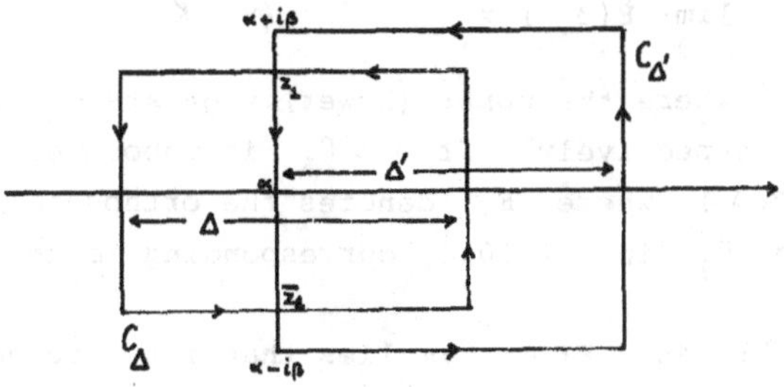

C_Δ , $C_{\Delta'}$ as indicated and denote by $C_\Delta^{\Delta'}$ the part of C_Δ which is inside $C_{\Delta'}$. At the points $z_1, \bar{z}_1 \in C_\Delta \cap C_{\Delta'}$ we take the Cauchy principal value of the integral along C_Δ , which will be denoted by a second prime at the integral. Then, applying the resolvent relation, we get

$$\tilde{E}(\Delta)\,\tilde{E}(\Delta') = -\frac{1}{4\pi^2}\int''_{C_\Delta} R_z\Big(\int'_{C_{\Delta'}} R_\zeta\, x\, d\zeta\Big)\, dz =$$

$$= -\frac{1}{4\pi^2}\int''_{C_\Delta} R_z \int'_{C_{\Delta'}} (z-\zeta)^{-1}\, d\zeta\, x\, dz + \frac{1}{4\pi^2}\int''_{C_\Delta}\Big(\int'_{C_{\Delta'}} (z-\zeta)^{-1} R_\zeta\, x\, d\zeta\Big) dz =$$

$$= -\frac{1}{2\pi i}\int_{C_\Delta^{\Delta'}} R_z x\, dz - \frac{1}{2\pi i}\int_{C_{\Delta'}^{\Delta}} R_\zeta\, x\, d\zeta = -\frac{1}{2\pi i}\int_{C_{\Delta\cap\Delta'}} R_z\, x\, dz = \tilde{E}(\Delta\cap\Delta') .$$

Here, in order to get the fourth equality sign we apply Cauchy's formula:

$$-\frac{1}{2\pi i}\int'_{C_{\Delta'}} (z-\zeta)^{-1}\, d\zeta = \begin{cases} 1 & \text{if } z \text{ is inside } C_{\Delta'} \\ 0 & \text{if } z \text{ is outside } C_{\Delta'} \end{cases}$$

and write for sufficiently small $\delta > 0$:

$$\int'_{C_{\Delta'}} (z-\zeta)^{-1} R_\zeta\, x\, d\zeta = i\int'^{\,\delta}_{-\delta} (z-\alpha-it)^{-1} R_{\alpha+it}\, x\, dt +$$

$$+ \int_{C_{\Delta'}\setminus\{\alpha+is:-\beta\le s\le\beta\}} (z-\zeta)^{-1} R_\zeta\, x\, d\zeta + i\Big(\int_\beta^\delta + \int_{-\delta}^{-\beta}\Big)(z-\alpha-it)^{-1} R_{\alpha+it}\, x\, dt .$$

Now the integration along C_Δ can be done without difficulty using the Remark at the end of section 1. If $\Delta\subset\Delta'$ or $\Delta\cap\Delta' = \emptyset$, the relation (3.2) follows in the same way, however, with considerable simplifications.

For a bounded interval $\Delta \in R_A$ with endpoints $\alpha_1 \le \alpha_2$ we define the operator $E(\Delta)$ as follows:

$$E(\Delta)x := \lim_{\varepsilon \downarrow 0} \tilde{E}(\Delta_\varepsilon)\, x \qquad (x \in K)$$

with $\Delta_\varepsilon := [\alpha_1 \mp \varepsilon, \alpha_2 \pm \varepsilon]$, where the upper (lower) sign at α_j is chosen if $\alpha_j \in \Delta$ ($\alpha_j \notin \Delta$, respectively). If $\Delta \in R_A$ is unbounded we put $E(\Delta) := E_1 - E(R^1 \setminus \Delta)$, where E_1 denotes the orthogonal projector onto the subspace K_1 in (2.10), corresponding to the real spectrum of A.

The definition of $\tilde{E}(\Delta)$ and $E(\Delta)$ implies that for a bounded interval Δ we have

$$\tilde{E}(\Delta) = E(\Delta) = 0 \qquad \text{if} \quad \bar{\Delta} \cap \sigma(A) = \emptyset \tag{3.3}$$

The property (1) follows from the relation $\tilde{E}(\Delta) = \tilde{E}(\Delta)^+$, which is a consequence of the symmetry of C_Δ with respect to the real axis. If $\Delta, \Delta' \in R_A$ are bounded and we choose sequences $(\varepsilon_j), (\varepsilon_j')$, tending to zero and such that the intervals Δ_{ε_j}, $\Delta'_{\varepsilon_j'}$ have no common endpoints, we get from (3.2)

$$E(\Delta)E(\Delta')\, x = \lim_{j\to\infty} \tilde{E}(\Delta_{\varepsilon_j})\, \tilde{E}(\Delta'_{\varepsilon_j'})\, x =$$

$$= \lim_{j\to\infty} \tilde{E}(\Delta_{\varepsilon_j} \cap \Delta'_{\varepsilon_j'})x = E(\Delta \cap \Delta')x \quad (x \in K).$$

In a similar way, (3) follows from (3.1) if we only choose the approximating closed intervals Δ_{ε_j}, $\Delta'_{\varepsilon_j'}$ such that $\Delta_{\varepsilon_j} \cap \Delta'_{\varepsilon_j'} = \emptyset$, $\Delta_{\varepsilon_j} \cup \Delta'_{\varepsilon_j'} \in R_A$. Now it is easy to extend (2) and (3) to unbounded elements of R_A.

Let $\Delta \in R_A$ be bounded and such that, e.g., $p > 0$ on $\bar{\Delta} \cap \sigma(A)$. We can additionally suppose that p has no zeros on $\bar{\Delta}$; otherwise we choose a suitable decomposition of Δ and observe (3.3). Then the Corollary 2 of Lemma 1.1 implies

$$[E(\Delta)x, x] = \int_{\alpha_1 \mp 0}^{\alpha_2 \pm 0} q(t)^{-1}\, d[JF(t)\, x, x] \quad (x \in K), \tag{3.4}$$

where, again, the upper (lower) sign at α_j has to be chosen if $\alpha_j \in \Delta$ ($\alpha_j \notin \Delta$, respectively). Thus (4) follows for bounded $\Delta \in R_A$. The property (5) is obvious from the definition of $E(\Delta)$.

Let again $\Delta \in R_A$ be bounded and z be outside of C_Δ. Then we have

$$R_z \tilde{E}(\Delta) x = -\frac{1}{2\pi i} \int_{C_\Delta} R_z R_\zeta x \, d\zeta = -\frac{1}{2\pi i} \int_{C_\Delta} (z-\zeta)^{-1}(R_z - R_\zeta) x \, d\zeta =$$

$$= -\frac{1}{2\pi i} \int_{C_\Delta} (\zeta - z)^{-1} R_\zeta x \, d\zeta$$

and this expression is holomorphic if z is outside C_Δ. As C_Δ can be chosen arbitrarily close to the real axis, the inclusion (7) follows for bounded Δ. Now Corollary 2 of Proposition 2.1 implies that $A|_{E(\Delta)K}$ is bounded.

If in representation (2.9) we replace F by E according to (3.4), it follows that

$$R_z = q(z)^{-1}\{ \int_{-\infty}^{\infty} q(t)(t-z)^{-1} dE(t) + \sum_{\alpha \in N(p) \cap R^1} (\alpha - z)^{-1}(F(\alpha+0) - F(\alpha-0)) - Q(z,A)\} . \tag{3.5}$$

Here the integral is to be understood as an improper integral with respect to the possible singularities of E at the zeros of p ; it converges in the s.o.t.

Now let $\Delta \in R_A$ be unbounded. Then $\Delta' := R^1 \setminus \Delta$ is a bounded element of R_A, and the inclusion in (7) follows if we show that $E(\Delta')x = 0$ implies the holomorphy of $R_z x$ in the inner points of Δ'. If Δ' does not contain any zero of p this follows immediately from (3.5) and (2). If Δ' contains some zero α of p in its interior then, by (3.5), $R_z x$ is holomorphic or has a pole at α. In the second case for an arbitrary small interval $\Delta'' \in R_A$ around α we would get

$$0 \neq E(\Delta'')x = E(\Delta'') E(\Delta') x = 0 ,$$

a contradiction. Thus (7) is proved.

It remains to prove (4) for unbounded $\Delta \in R_A$. This will be done in section 5.

REMARK. The homomorphism

$$\Delta \to E(\Delta) \qquad (\Delta \in R_A) \tag{3.6}$$

can be extended to a much larger class of subsets of the real axis. In order to see this we have only to observe that for bounded intervals $\Delta \in R_A$ with $\Delta \cap N(p) = \emptyset$ one of the spaces $(E(\Delta)K, \pm[\cdot,\cdot])$ is a Hilbert space (see Remark 1 after Theorem I.5.2) and to apply well-known results for spectral functions in Hilbert spaces. E.g.,

(3.6) can be extended to all bounded Borel sets whose boundary points have a positive distance from $c(A)$.

Let again $\Delta_o \in R_A$ be bounded and such that $\Delta_o \cap N(p) = \emptyset$ for some definitizing polynomial p of A. Assume, e.g., that $p > 0$ on Δ_o. Then $(E(\Delta_o)K, [\cdot,\cdot])$ is a Hilbert space and $A|_{E(\Delta_o)K}$ is a bounded selfadjoint operator in this Hilbert space. Evidently, $E(\Delta \cap \Delta_o)$ is the spectral function of $A|_{E(\Delta_o)K}$. This observation easily extends to the following

COROLLARY. _Let_ $\Delta_o \in R_A$ _be bounded and such that_ $\Delta_o \cap c(A) = \emptyset$. _Then we have_

$$A\,E(\Delta_o) = \int_{\Delta_o} t\, d\,E(t)\,. \tag{3.7}$$

The representations given in the following theorem have been established already in the proof of Theorem 3.1.

THEOREM 3.2. _If_ A _is a definitizable operator in_ K _with definitizing polynomial_ p _the following representations hold_ $(q(z) = =p(z)(z-z_o)^{-k}(z-\bar{z}_o)^{-k})$:

$$q(A)\,R_z = \int_{R^1} q(t)(t-z)^{-1}\,d\,E(t) + \sum_{\alpha\in N(p)\cap R^1} (\alpha-z)^{-1} N_\alpha\,, \tag{3.8}$$

$$q(A) = \int_{R^1} q(t)\,d\,E(t) + N\,. \tag{3.9}$$

Here the integrals are improper ones with respect to the points $\alpha \in N(p)\cap R^1$ _and_ ∞ ; _they converge in the s.o.t. The operators_ N_α $(\alpha \in N(p)\cap R^1)$ _have the properties_

$$[N_\alpha x,x] \geq 0 \;(x \in K),\; N_\alpha E(\Delta) = 0 \;(\Delta \in R_A,\ \alpha \notin \bar{\Delta})$$

$$N_\alpha N_{\alpha'} = 0\,,\; (A - \alpha I)\,N_\alpha = 0$$

_if_ $\alpha, \alpha' \in N(p)\cap R^1$, _and_ $N := \sum_{\alpha\in N(p)\cap R^1} N_\alpha$.

In order to obtain (3.8) and (3.9) we only have to replace in (2.8) F by E (see (3.4)) and to put $N_\alpha := J(F(\alpha+0)-F(\alpha-0))$ $(\alpha \in N(p)\cap R^1)$. It will be left to the reader to verify the properties of the operators N_α .

COROLLARY. *If* $\alpha \in c(A)$ *, the integral*

$$\int_{\alpha-\delta}^{\alpha+\delta} (t-\alpha)^{k(\alpha)}\, d\,E(t)$$

converges for sufficiently small $\delta > 0$ *in the s.o.t. Here* $k(\alpha)$ *denotes again the order of the zero* α *of a definitizing polynomial of* A .

II.4. *A characterization of the critical points.* We start with the following

LEMMA 4.1. *Let* A *be a definitizable operator in* K *and* p *be a definitizing polynomial of* A *. If* $\beta \in \rho(A)$ *is a real zero of* p *of multiplicity* $k(\beta)(>0)$ *and* $\hat{\beta}$ *is an arbitrary point of the same (connected) open component of* $\rho(A) \cap R^1$ *as* β *, then*

$$\hat{p}(\lambda) := (\lambda-\hat{\beta})^{k(\beta)} (\lambda-\beta)^{-k(\beta)} p(\lambda)$$

is also a definitizing polynomial of A .

PROOF. The representation (3.8) implies

$$q(A)\, R_{\beta}^{k(\beta)} = \int_{R^1} q(t)(t-\beta)^{-k(\beta)} d\,E(t) + \sum_{\alpha \in N(p) \cap R^1} (\alpha-\beta)^{-k(\beta)} N_{\alpha} .$$

Observing (3.7) , we find

$$q(A)(A-\hat{\beta} I)^{k(\beta)} R_{\beta}^{k(\beta)} = \int_{R^1} q(t)(t-\hat{\beta})^{k(\beta)} (t-\beta)^{-k(\beta)} d\,E(t) +$$

$$+ \sum_{\alpha \in N(p) \cap R^1} (\alpha-\hat{\beta})^{k(\beta)} (\alpha-\beta)^{-k(\beta)} N_{\alpha} = \int_{R^1} (t-\hat{\beta})^{k(\beta)} (t-\beta)^{-k(\beta)} J\, d\,F(t),$$

and the statement follows as $(t-\hat{\beta})(t-\beta)^{-1}$ is positive on $\sigma(A) \cap R^1$.

The critical points of the definitizable operator A can be characterized as follows.

PROPOSITION 4.2. *We have* $\alpha \in c(A)$ *if and only if for each* $\Delta \in R_A$ *with* $\alpha \in \Delta$ *the scalar product* $[\cdot,\cdot]$ *is indefinite on* $E(\Delta)K$.

PROOF. If $\alpha \notin c(A)$ then $\alpha \notin \sigma(A)$ or $\alpha \notin N(p)$ for some definitizing polynomial p of A . In the first case $\Delta \in R_A$ can be chosen such that $\overline{\Delta} \cap \sigma(A) = \emptyset$, $\alpha \in \Delta$, hence $E(\Delta) = 0$. In the second case $\Delta \in R_A$ can be chosen such that p is of one sign on $\overline{\Delta} \cap \sigma(A)$. Thus, in both cases on $E(\Delta)K$ the s.p. $[\cdot,\cdot]$ is not indefinite.

Now let $\alpha \in c(A)$ and assume that, e.g. $E(\Delta)K$ is a positive subspace for some bounded $\Delta \in R_A$, $\alpha \in \Delta$. Consider the decomposition

$$K = E(\Delta)K \dot{+} (I-E(\Delta))K . \tag{4.1}$$

If α is a zero of odd order for some definitizing polynomial p of A, then according to Theorem 3.1, (4), on one side of α there are no points of $\sigma(A) \cap \Delta$. If $\hat{\alpha} \in \Delta, \hat{\alpha} \neq \alpha$, is on this side, we have with $\hat{p}(\lambda) = (\lambda - \hat{\alpha})(\lambda - \alpha)^{-k(\alpha)} p(\lambda)$:

$$[\hat{p}(A)x,x] \geqq 0 \qquad (x \in E(\Delta)K).$$

This follows from the fact that $\hat{p}(\lambda) > 0$ if $\lambda \in \sigma(A) \cap \Delta$ and that $(E(\Delta)K, [\cdot,\cdot])$ is a Hilbert space. If α is a zero of even order of p and we choose $\hat{\alpha} \in \Delta$, $\hat{\alpha} \neq \alpha$, then by the same reasoning, $[\hat{p}(A)x,x] \geqq 0 (x \in E(\Delta)K)$.

It remains to consider the space $(I-E(\Delta))K$. According to Lemma 4.1, the zero α of p can be replaced by $\hat{\alpha}$ chosen above, that is, we have also $[\hat{p}(A)x,x] \geqq 0$ $(x \in (I-E(\Delta))K \cap \mathcal{D}(A^k))$. Therefore $\hat{p}$ is a definitizing polynomial of A and $\hat{p}(\alpha) \neq 0$, hence α cannot belong to $c(A)$. The proposition is proved.

By $\sigma_+(A)$ $(\sigma_-(A))$ we denote the set of all $\lambda \in \sigma_1(A) = \sigma(A) \cap R^1$ for which there exists an interval $\Delta \in R_A$, $\lambda \in \Delta$, such that $E(\Delta)K$ is a positive (negative, respectively) subspace of K. The Proposition 4.2 implies that the real spectrum $\sigma_1(A)$ decomposes as

$$\sigma_1(A) = c(A) \cup \sigma_+(A) \cup \sigma_-(A) .$$

The points of $\sigma_+(A)$ $(\sigma_-(A))$ are called <u>spectral points of positive (negative, respectively) type</u> of A.

II.5. <u>Further study of critical points</u>. In this section A is again a definitizable operator with definitizing polynomial p and spectral function E, see Theorem 3.1. We shall study the behavior of E in a neighbourhood of a critical point $\alpha \in c(A)$ and at ∞. To this end, if $\alpha \in c(A)$ we put

$$S_\alpha := \bigcap_{\Delta \in R_A,\ \alpha \in \Delta} E(\Delta)K .$$

PROPOSITION 5.1. S_α is the algebraic eigenspace of A corresponding to the eigenvalue α. Moreover, we have $(A-\alpha I)^{k(\alpha)+1} S_\alpha = \{0\}$.

PROOF. If $x \in E(\Delta)K$ the singularities of the function $z \to R_z x$ are contained in $\bar{\Delta}$. Thus, if $x \in S_\alpha$, the only singularities of this function are at $z = \alpha$, and conversely. Now the statement follows from Proposition 2.1.

Without loss of generality we can often suppose in the following that α is the only critical point of A, that A is bounded and that the definitizing polynomial is of the form $p(\lambda) = \pm(\lambda-\alpha)^k$. Indeed, in order to study E in a neighbourhood of α we consider the restriction of A or E to the subspace $E(\Delta)K$ where $\Delta \in R_A$ is a bounded interval which contains α but no other zeros of p. Then we have $p(\lambda) = (\lambda-\alpha)^{k(\alpha)} p_o(\lambda)$ with a polynomial p_o which is positive or negative on Δ. The operator $p_o(A)^{-1/2}$ ($(-p_o(A))^{-1/2}$, respectively) can be defined by the Riesz-Dunford functional calculus and we get

$$\pm[(A-\alpha I)^{k(\alpha)}x,x] = [(A-\alpha I)^{k(\alpha)} p_o(A)\, y,y] \geq 0$$

$$(x \in E(\Delta)K,\ y := (\pm p_o(A))^{-1/2}x) .$$

If the exponent k of the definitizing polynomial $\pm(\lambda-\alpha)^k$ is even, all the points of $\sigma(A) \setminus \{\alpha\}$ are of the same (positive or negative) type ; if k is odd, points of $\sigma(A) \setminus \{\alpha\}$ on opposite sides of α are of opposite type. We put

$$S_\alpha^{\pm} := \text{c.l.s.}\{E(\Delta)K : \Delta \in R_A,\ \Delta \cap \sigma(A) \subset \sigma_\pm(A)\} . \tag{5.1}$$

PROPOSITION 5.2. Let A be a bounded definitizable operator in K with definitizing polynomial $p(\lambda) = \pm(\lambda-\alpha)^k$. Then the following statements hold true:

(1) $S_\alpha^{\pm} \subset P_\pm$.

(2) The subspaces $S_\alpha, S_\alpha^{+}, S_\alpha^{-}$ are mutually orthogonal, invariant under A and all operators $E(\Delta)$ $(\Delta \in R_A)$ and we have $\sigma(A\,|S_\alpha^{\pm}) = \overline{\sigma_\pm(A)}$.

(3) $S_\alpha = (S_\alpha^+ + S_\alpha^-)^\perp$.

PROOF. (1) is an easy consequence of Lemma I.5.3. The statement (2) follows from the definitions of S_α, $S_\alpha^\pm$ and Theorem 3.1. Finally, the orthogonality of S_α and $S_\alpha^+ + S_\alpha^-$ implies $S_\alpha \subset (S_\alpha^+ + S_\alpha^-)^\perp$. Conversely, if $x \perp (S_\alpha^+ + S_\alpha^-)$ we have $x \perp E(\Delta)K$ for all intervals $\Delta \in R_A$ not containing α, that is, $E(\Delta')x = x$ for all $\Delta' \in R_A$, $\alpha \in \Delta'$, hence $x \in S_\alpha$.

COROLLARY 1. $S_\alpha^+ \cap S_\alpha^- \subset S_\alpha$.

COROLLARY 2. <u>The relations</u>

$$S_\alpha = \{0\} \text{ \underline{and} } \overline{S_\alpha^+ + S_\alpha^-} = K$$

<u>are equivalent.</u>

COROLLARY 3. <u>If</u> $\alpha \in c(A)$ <u>and</u> $S_\alpha = \{0\}$, <u>then one of the polynomials</u> $p(\lambda) = \pm(\lambda - \alpha)$ <u>is definitizing for</u> A .

Indeed, assume, e.g. that $\sigma_-(A) < \alpha$ and $\sigma_+(A) > \alpha$. Then we have for $x = x_+ + x_-$ with $x_\pm \in$ l.s. $\{E(\Delta)K : \Delta \in R_A,\ \Delta \cap \sigma(A) \subset \sigma_\pm(A)\}$:

$$[(A-\alpha I)x,x] = [(A-\alpha I)x_+,x_+] + [(A-\alpha I)x_-,x_-] \geq 0$$

as both the s.p.-s in the middle term are nonnegative. The set of these x is dense in $S_\alpha^+ \dot{+} S_\alpha^-$ and hence also dense in K.

Among all the definitizing polynomials of a definitizable operator A there exist such of <u>minimal degree</u>. It is easy to see, that for a definitizing polynomial p of minimal degree each nonreal zero belongs to $\sigma_p(A)$. Moreover, a real zero α of such a polynomial p which belongs to $\sigma_+(A) \cup \sigma_-(A) \cup \rho(A)$ is of first order. Indeed, if e.g. $\alpha \in \rho(A)$, assume $p(\lambda) = (\lambda - \alpha)^{2\kappa} p_o(\lambda)$ with some integer $\kappa > 0$, where either $p_o(\alpha) \neq 0$ or α is a simple zero of p_o . It follows

$$[p_o(A)x,x] = [p(A)R_\alpha^\kappa x, R_\alpha^\kappa x] \geq 0 \qquad (x \in \mathcal{D}(A^{k_o+\kappa})) ,$$

$(k_o := \deg p_o)$, that is

$$[p_o(A) R_\alpha^{k_o+\kappa} x, R_\alpha^{k_o+\kappa} x] \geq 0 \qquad (x \in K) .$$

As $R_\alpha^\kappa K$ is dense in K and $p_o(A) R_\alpha^{k_o}$ is bounded we get

$$[p_o(A) R_\alpha^{k_o} x, R_\alpha^{k_o} x] \geq 0 \qquad (x \in K) .$$

Hence p was not of minimal degree. The cases $\alpha \in \sigma_{\pm}(A)$ can be treated similarly (comp. the proof of Proposition 4.2).

The Corollary 3 yields the following conclusion for an arbitrary definitizable operator A.

PROPOSITION 5.3. <u>If</u> A <u>is a definitizable operator in</u> K, $\alpha \in c(A)$ <u>and</u> $S_\alpha = \{0\}$, <u>each definitizing polynomial of minimal degree has a simle zero at</u> α.

If one of the spaces S_α^+, S_α^- consists only of 0, the Corollary 2 implies $S_\alpha \neq \{0\}$. This yields the following

PROPOSITION 5.4. <u>If</u> A <u>is an arbitrary definitizable operator in</u> K, $\alpha \in c(A)$ <u>and there exists a neighbourhood</u> Δ <u>of</u> α <u>such that all the points of</u> $(\Delta \cap \sigma(A)) \setminus \{\alpha\}$ <u>are of the same type</u> (this holds e.g. <u>if</u> α <u>is a zero of even order of some definitizing polynomial</u> p <u>of</u> A) <u>then</u> $\alpha \in \sigma_p(A)$.

Let A be again a definitizable operator in K. We extend the set $c(A)$ as follows:

$\tilde{c}(A) := c(A) \cup \{\infty\}$, if for each unbounded set $\Delta \in R_A$ the space $E(\Delta)K$ is indefinite; otherwise we put $\tilde{c}(A) := c(A)$.

If ∞ belongs to $\tilde{c}(A)$ it will be called a <u>critical point</u> of A. It is easy to see that this holds if and only if in each neighbourhood of ∞ there are points of $\sigma_+(A)$ as well as of $\sigma_-(A)$. Indeed, suppose that in some neighbourhood $\Delta' \in R_A$ of ∞ there are points of at most one type of $\sigma(A)$. We can assume that $0 \notin \overline{\Delta'}$, and consider the operator $A' := A|_{E(\Delta')K}$ in $E(\Delta')K$. Then $(A')^{-1}$ is a definitizable operator with the only possible critical point 0, and all the points of $\{\lambda^{-1} : \lambda \in \Delta'\} \cap \sigma((A')^{-1})$ are of the same type. Then, according to Proposition 5.4, $0 \in \sigma_p((A')^{-1})$, which is impossible. This consideration proves also the statement (4) of Theorem 3.1 for unbounded Δ.

The same reasoning implies that for an arbitrary definitizable operator A we have

$$S_\infty := \bigcap_{\Delta \in R_A,\ \Delta \text{ unbounded}} E(\Delta)K = \{0\}.$$

If $\alpha \notin \tilde{c}(A)$, for arbitrary $\lambda_0, \lambda_1 \in R^1 \setminus c(A)$, $\lambda_0 < \alpha$, $\lambda_1 > \alpha$, the limits

$$\lim_{\lambda \uparrow \alpha} E([\lambda_0, \lambda]), \quad \lim_{\lambda \downarrow \alpha} E([\lambda, \lambda_1]) \tag{5.2}$$

exist in the s.o.t. Here we agree that, if $\alpha = \infty$, then $\lambda_1 > \alpha$ $(\lambda \downarrow \alpha)$

means $\lambda_1 > -\infty$ ($\lambda \downarrow -\infty$, respectively). If $\alpha \in \tilde{c}(A)$ and the limits (5.2) do still exist, then α is called a <u>regular critical point</u> of A or E, otherwise α is called <u>singular.</u>

In Lemma 5.5. and Proposition 5.6. we suppose that $\tilde{c}(A) = \{\alpha\}$ and that a definitizing polynomial is of the form

$$p(\lambda) = (\lambda - \alpha)^k \quad \text{or} \quad p(\lambda) = \lambda^k$$

if $\alpha \neq \infty$ or $\alpha = \infty$, respectively. The sets $S_\infty^\pm$ are also defined by the relation (5.1). Moreover, we choose an (arbitrary) increasing sequence (Δ_n^+) of sets $\Delta_n^+ \in R_A$, such that $\alpha \notin \Delta_n^+$, $\bigcup_1^\infty \Delta_n^+ \supset \sigma_+(A)$ and $(\bigcup \Delta_n^+) \cap \sigma_-(A) = \emptyset$. It is easy to see that

$$S_\alpha^+ = \overline{\bigcup_1^\infty E(\Delta_n^+)K} .$$

LEMMA 5.5. <u>If</u> $x \notin S_\alpha^+ + (S_\alpha^+)^\perp$ <u>we have</u>

$$\|E(\Delta_n^+)x\| \to \infty \qquad (n \to \infty) .$$

PROOF. Consider $x \in K$ such that $\sup_n \|E(\Delta_n^+)x\| < \infty$. Then the sequence $(E(\Delta_n^+)x)$ contains a weakly convergent subsequence, and we can suppose that $E(\Delta_n^+)x \to x_0$ (weakly) if $n \to \infty$. The sequence $(x - E(\Delta_n^+)x)$ is also bounded and we can suppose that it converges: $x - E(\Delta_n^+)x \to y_0$ (weakly) if $n \to \infty$. Then for arbitrary $k = 1,2,\ldots$

$$E(\Delta_k^+)y_0 = \lim_{n\to\infty} E(\Delta_k^+)(x - E(\Delta_n^+)x) = 0 ,$$

hence $y_0 \in (S_\alpha^+)^\perp$, and we finally get

$$x = \lim_{n\to\infty}(E(\Delta_n^+)x + (x - E(\Delta_n^+)x)) = x_0 + y_0 \in S_\alpha^+ + (S_\alpha^+)^\perp .$$

PROPOSITION 5.6. <u>The following statements are equivalent:</u>

(i) $\alpha \in \tilde{c}(A)$ <u>is a regular critical point.</u>

(ii) $K = S_\alpha \dot{+} S_\alpha^+ \dot{+} S_\alpha^-$ [1)]

(iii) <u>It exists a number</u> k <u>such that</u>

$$\|E(\Delta)\| \leq k \quad \underline{\text{for all}} \quad \Delta \in R_A .$$

---

1) If $\alpha = \infty$ then $S_\infty = \{0\}$.

PROOF. (i) => (ii) : Assume that $(S_\alpha^+, [\cdot,\cdot])$ is not a Hilbert space. This implies $K \neq S_\alpha^+ + (S_\alpha^+)^\perp$. Then, if we consider $x \in K \setminus (S_\alpha^+ + (S_\alpha^+)^\perp)$, it follows from Lemma 5.5 that α cannot be regular. Thus $(S_\alpha^+, [\cdot,\cdot])$ and, in the same way, $(S_\alpha^-, -[\cdot,\cdot])$ are Hilbert spaces. Then $S_\alpha^+ + S_\alpha^-$ is a Krein space, $S_\alpha^+ \cap S_\alpha^- = \{0\}$, and (ii) follows from Proposition 5.2,(3).

(ii)=> (iii): Let $x = x_0 + x_1 + x_2$ be the decomposition of $x \in K$ according to (ii). As $(S_\alpha^+, [\cdot,\cdot])$ and $(S_\alpha^-, -[\cdot,\cdot])$ are Hilbert spaces, there exist positive constants $\gamma_j', \gamma_j'', \beta_\ell$ (j=1,2; ℓ = 0,1,2) such that

$$\gamma_j' \| x_j \| \leq |[x_j, x_j]|^{1/2} \leq \gamma_j'' \| x_j \| \quad (j=1,2),$$

$$\| x_\ell \| \leq \beta_\ell \| x \| \quad (\ell=0,1,2).$$

Further, S_α, $S_\alpha^\pm$ are invariant with respect to all operators $E(\Delta)$ $(\Delta \in R_A)$, whence

$$\|E(\Delta)x\| \leq \|E(\Delta)x_1\| + \|E(\Delta)x_2\| + \|x_0\| \leq$$

$$\leq \frac{1}{\gamma_1'} [E(\Delta)x_1, x_1]^{1/2} + \frac{1}{\gamma_2'} |[E(\Delta)x_2, x_2]|^{1/2} + \|x_0\| \leq$$

$$\leq \left(\frac{\gamma_1''}{\gamma_1'} \beta_1 + \frac{\gamma_2''}{\gamma_2'} \beta_2 + \beta_0\right) \| x \| .$$

Finally, (iii) implies also (i) according to a Theorem of E.R.Lorch [28].

The following theorem is now an easy consequence of Proposition 5.6.

THEOREM 5.7. Let A be a definitizable operator in the Krein space K and $\alpha \in \tilde{c}(A)$. The following statements are equivalent:

(i) α is a regular critical point of A.

(ii) For arbitrary $\Delta \in R_A$, $\alpha \in \Delta$[2)], there exist orthogonal and mutually orthogonal projectors $E_-(\Delta)$, $E_+(\Delta)$ and E_α in the double commutant of R_z $(z \in \rho(A))$ such that

---

2) If $\alpha = \infty$ we put $\alpha \in \Delta$ if Δ is unbounded.

$$E(\Delta) = E_+(\Delta) + E_-(\Delta) + E_\alpha .$$

$$\sigma(A\,|E_\mp(\Delta)K) \subset \overline{\Delta \cap \{\lambda : \lambda \lessgtr \alpha\}}$$ and $E_\alpha K = S_\alpha$.

(iii) For each neighbourhood U of α with $U \cap \tilde{c}(A) = \{\alpha\}$ there exists a $k > 0$ such that

$$\|E(\Delta)\| \leq k \quad \text{for all } \Delta \in R_A,\ \Delta \subset U .$$

Theorem 5.7 implies that the homomorphism $\Delta \to E(\Delta)$ can be extended to the semiring generated by those bounded intervals whose endpoints are not singular critical points of A. For this extension the properties (1)-(7) of Theorem 3.1 are preserved.

Examples of definitizable operators with a singular critical point α and $S_\alpha = \{0\}$ were given in [29]. Criteria for the regularity of the critical points of a nonnegative operator are given in [30], [31].

In [5] (see also [13]) the reader can find a detailed study of those critical points $\alpha \in c(A)$, for which the space $E(\Delta)K$ is a π_κ-space for some $\Delta \in R_A, \alpha \in \Delta$.

II.6. Nonnegative operators. In this section we give some integral representations for nonnegative operators. In applications these representations yield, e.g., eigenfunction expansions for certain elements, see [4].

If the operator A in K is nonnegative and $\rho(A) \neq \emptyset$, it is definitizable with definitizing polynomial $p(\lambda) = \lambda$. Therefore, the only possible critical points are 0 and ∞. The statements of the following theorem are of interest only in the case that 0 and (or) ∞ are singular critical points. Otherwise the considerations in section 5 yield stronger results.

THEOREM 6.1. If A is a nonnegative operator in the Krein space K, $\rho(A) \neq \emptyset$, then the following representations hold:

(i) $$A x = \int_{R^1} t \, d E(t)\, x + N x \qquad (x \in \mathcal{D}(A^2)), \qquad (6.1)$$

(ii) $$x = \int_{R^1} d E(t)\, x + x_0 \qquad (x \in \mathcal{D}(A) \cap \mathcal{R}(A)),$$

(iii) $$(A - zI)^{-1} x = \frac{1}{z} \int_{R^1} (t-z)^{-1}\, t \, d E(t) x - z^{-2} N x - z^{-1} x \qquad (x \in \mathcal{D}(A)).$$

<u>The integrals are improper ones with respect to</u> 0 <u>and</u> ∞ <u>converging in the s.o.t., the operator</u> N <u>has the properties</u>

$$[\,Nx,x\,] \geqq 0 \quad (x \in K),\ N\,E(\Delta) = 0 \quad (\Delta \in R_A,\ 0 \notin \Delta)\ ,$$

$$N^2 = AN = 0\ ,$$

<u>and</u> $Ax_o = 0$.

PROOF. If we choose $q(\lambda) = \lambda(\lambda^2+1)^{-1}$ then (3.9) implies

$$A(A^2+I)^{-1}y = \int_{R^1} t(t^2+1)^{-1}\,d\,E(t)y + Ny \qquad (y \in K).$$

Putting $x: = (A^2+I)^{-1}y$ and observing that for $t \neq 0$ we have

$$(t^2+1)^{-1}d\,E(t)y = d\,E(t)\,x\ , \quad Ny = N(A^2+I)\,x = Nx\ ,$$

the representation (6.1) follows.

Before we continue the proof of Theorem 6.1, we observe the following statements which are, perhaps, of interest in their own. Here $\Delta_o \in R_A$ is a bounded interval containing 0 , $\Delta_\infty \in R_A$ is unbounded and does not contain 0 .

LEMMA 6.2. <u>The following integrals converge in the norm topology of</u> K :

(a) $\int_{\Delta_o} t\,d\,E(t)\,x$ if $x \in K$,

(b) $\int_{\Delta_\infty} t^{-1}d\,E(t)u$ if $u \in K$,

(c) $\int_{\Delta_o} d\,E(t)\,y$ if $y \in R(A)$,

(d) $\int_{\Delta_\infty} d\,E(t)\,v$ if $v \in D(A)$.

<u>The integral in</u> (d) <u>is equal to</u> $E(\Delta_\infty)v$, <u>the integral in</u> (c) <u>is equal to</u> $E(\Delta_o)y$ <u>if the kernel of</u> A <u>is</u> $\{0\}$.

Indeed, (a) follows from (i) in Theorem 6.1 which has already been proved. Relation (c) is obtained from (a) if we put $y = Ax$ and observe that $d\,E(t)y = t\,d\,E(t)x$ $(t \neq 0)$. If $0 \notin \sigma_p(A)$, then $N = 0$ in (i) and we get $E(\Delta_o)y = \int_{\Delta_o} dE(t)y$. In order to prove (b) we consider the operator $A' := (A|_{E(\Delta_\infty)K})^{-1}$ in $E(\Delta_\infty)K$. Then A' is bounded and nonnegative, hence $\int_{\Delta_\infty^{-1}} s\,d\,E'(s)\,u$ exists for arbitrary $u \in E(\Delta_\infty)K$ and equals $A'u$: here E' denotes the spectral function of A' . As $d\,E'(s) = d\,E(t)$ if $s = t^{-1}$, the statement follows. Finally, (d) is easily derived from (b) or (c).

Now we complete the proof of Theorem 6.1. The relation (ii) follows from (c), (d) and the last statement of the lemma. In order to get (iii) we first observe that (3.8) implies (comp. the proof of (i))

$$A(A-zI)^{-1}x = \int_{R^1} (t-z)^{-1}\, t\, d\, E(t)x - z^{-1}N\, x \quad (x \in \mathcal{D}(A^2))$$

which can be written as

$$(A-zI)^{-1}x = z^{-1} \int_{R^1} (t-z)^{-1} t\, d\, E(t)\, x - z^{-2}N\, x - z^{-1}x \quad (x \in \mathcal{D}(A^2)).$$

Approximating $x \in \mathcal{D}(A)$ by the sequence $(E([-n,n])x)$, $n = 1,2,\ldots,$ the relation (iii) follows.

II.7. Maximal nonnegative invariant subspaces. As an application of the spectral function we prove the following theorem. It is a special case of the results contained in [9] ; however, the proof in this case simplifies considerably.

THEOREM 7.1. Let A be a bounded nonnegative operator in the Krein space K. Then there exist a maximal nonnegative and a maximal nonpositive subspace which are invariant under A.

PROOF. We consider the subspaces $S_0^{\pm}$ introduced in section 5 and the range $N := NK$, where N is given by (6.1). Then the spaces S_0^+, S_0^- and N are mutually orthogonal and

$$\overline{S_0^+ + N} \subset P_+ , \quad S_0^- \subset P_- .$$

The representation (6.1) imples

$$A(S_0^-)^{\perp} \subset S_0^+ + N .$$

Indeed, if $x \perp S_0^-$ we have

$$A\, x = \int_0^{\infty} t\, d\, E(t)\, x + N\, x \in S_0^+ + N .$$

According to a result of R.S.Phillips (see [1]) the alternating pair of subspaces $\{ S_0^+ + N ,\ S_0^-\}$ (for this terminology see [1]) is contained in a maximal alternating pair $\{L_+, L_-\}$. Then we have $L_+ = L_-^{\perp} \subset (S_0^-)^{\perp}$ and

$$AL_+ \subset A(S_0^-)^{\perp} \subset S_0^+ + N \subset L_+ .$$

The theorem is proved.

REFERENCES

[1] J.Bognár, Indefinite Inner Product Spaces, Springer-Verlag, Berlin-Heidelberg-New York, 1974.

[2] T.Ando, Linear Operators in Krein spaces, Lecture Notes, Hokkaido University, Sapporo, 1979.

[3] M.G.Kreĭn, H.Langer, Some propositions on analytic matrix functions related to the theory of operators in the space Π_κ, Acta sci. math. 43 (1981), 181-205.

[4] K.Daho, H.Langer, Sturm-Liouville operators with an indefinite weight function, Proc. Roy. Soc. Edinburgh, 78 A (1977), 161-191.

[5] H.Langer, Spektraltheorie linearer Operatoren in J-Räumen und einige Anwendungen auf die Schar $L(\lambda) = \lambda^2 I + \lambda B + C$, Habilitationsschrift, Technische Universität Dresden, 1965.

[6] M.G.Kreĭn, Ju.L.Šmul'jan. J-polar representations of plus-operators Mat. Issled. 1,2 (1966), 172-210 [Russian].

[7] P.Jonas, On the functional calculus and the spectral function for definitizable operators in Krein space, Beiträge Anal. 16 (1981), 121-135.

[8] H.Langer, B.Najman, Perturbation theory for definitizable operators, J. Operator theory (to appear).

[9] H.Langer, Invariante Teilräume definisierbarer J-selbstadjungierter Operatoren, Ann. Acad. Sci. Fenn. Ser. A I, 475 (1971).

[10] M.G.Kreĭn, Introduction to the geometry of indefinite J-spaces and to the theory of operators in these spaces. In: Second mathematical summer school, part I, 19-92, Naukova Dumka, Kiev 1965 [Russian].

[11] T.Ja.Azizov, I.S.Iohvidov, Linear operators in spaces with indefinite metric and their spplications, Itogi Nauki i Tehniki, Ser.Mat.Analiz 17(1979), 113-205 [Russian].

[12] I.S.Iohvidov, M.G.Kreĭn, H.Langer, Introduction to the spectral theory of operators in spaces with an indefinite metric, Mathematical Research, vol.9. Akademie- Verlag, Berlin, 1982.

[13] M.G.Kreĭn, G.K.Langer, On the spectral function of a selfadjoint operator in a space with indefinite metric, Dokl. Akad. Nauk SSSR 152(1963), 39-42 [Russian].

[14] N.Dunford, J.T.Schwartz, Linear Operators,I,General theory, Interscience, New York-London, 1958.

[15] P.Jonas, H.Langer, Compact perturbations of definitizable operators, J.Operator Theory 2 (1979), 63-77.

[16] I.C.Gohberg, M.G.Kreĭn , Theory of Volterra operators in Hilbert space and its applications, Nauka, Moscow, 1967 [Russian] .

[17] V.I.Kogan, F.S.Rofe-Beketov, On square-integrable solutions of symmetric systems of differential equations of arbitrary order, Proc. Roy. Soc. Edinburgh, 74 A (1974/75),5-40.

[18] Ju.L.Daleckiĭ, M.G.Kreĭn, The stability of the solutions of differential equations in a Banach space, Nauka, Moscow 1972 [Russian].

[19] M.G.Kreĭn, On weighted integral equations the distribution functions of which are not monotonic. In: Memorial volume dedicated to D.A.Grave, 88-103, Gostehizdat, Moscow-Leningrad, 1040 [Russian] .

[20] M.G.Kreĭn, H.Langer, On some mathematical principles in the linear theory of damped oscillations of continua, Integral Equations Operator Theory 1(1978),364-399 and 539-566.

[21] A.S.Sil'bergleit, Ju.I.Kopilevič, On the properties of waves connected with quadratic operator pencils, Dokl. Akad. Nauk SSSR 256 (1981), 565-570 [Russian].

[22] A.S.Sil'bergleit, Ju.I.Kopilevič, Mathematical theory of wave guide systems connected with quadratic operator pencils, Academy of Sciences of the USSR, Phys.-tech. Institute A.I.Ioffe, Preprint, Leningrad, 1980 [Russian].

[23] H.Langer, Uber eine Klasse polynomialer Scharen selbadjungierter Operatoren im Hilbertraum, II, J.Funct.Anal.16, 2 (1974), 221-234 .

[24] H.Weyl, Uber gewöhnliche lineare Differentialgleichungen mit singulären Stellen und ihre Eigenfunktionen, Nachr. Akad. Wiss. Göttingen, Math.-Phys.Kl.(1910),442-467, and in H.Weyl, Gesammelte Abhandlingen,I,Springer-Verlag, Berlin-Heidelberg-New York, 1968.

[25] A.Dijksma, Eigenfunction expansions for a class of selfadjoint ordinary differential operators with boundary conditions containing the eigenvalue parameter, Proc. Roy Soc. Edinburgh, 86 A(1980), 1-27.

[26] J.A.Ball, W.Greenberg, A Pontrjagin space analysis of the supercritical transport equation, Transport Theory Statist. Phys. 4,4(1975), 143-154.

[27] I.S.Kac, M.G.Kreĭn, R-functions-analytic functions mapping the upper half plane into itself, Appendix I to the russian edition of F.V. Atkinson, Discrete and Continuous Boundary Problems, Mir, Moscow, 1968 [Russian] .

[28] E.R.Lorch, On a calculus of operators in reflexive vector spaces, Trans. Amer. Math. Soc. 45 (1939) ,217-234.

[29] H.Langer, On maximal dual pairs of invariant subspaces of J-selfadjoint operators, Mat. Zametki 7 (1970), 443-447 [Russian].

[30] K.Veselić, On spectral properties of a class of J-selfadjoint operators, I, Glasnik Mat. Ser III, 7 (1972),229-248.

[31] R.V.Akopjan, On the regularity at infinity of the spectral function of a J-nonnegative operator,Izv.Akad.Nauk Armjan. SSR Ser. Mat.15 (1980), 357-364 [Russian] .

SEMIGROUPS AND COSINE FUNCTIONS

Svetozar Kurepa

INTRODUCTION

Exponential and cosine functions are closely related with differential equations:

$$x'(t) = Ax(t) \quad , \quad x''(t) + B^2x(t) = 0$$

with corresponding initial conditions. Solutions are given by

$$E(t) = e^{tA} \quad , \quad C(t) = \cos Bt \ .$$

We can say that these functions are defined by the above equations, or if we know how to define functions E and C, then we can solve the above equations. In Calculus one defines functions E and C by use of the power series:

$$E(t) = \sum_{n=0}^{\infty} \frac{t^n A^n}{n!} \quad , \quad C(t) = \sum_{n=0}^{\infty} (-1)^n \frac{t^{2n}B^{2n}}{(2n)!} \quad .$$

By direct calculation one verifies that so defined functions satisfy the following functional equations:

(1) $$E(t+s) = E(t)\,E(s) \ , \ E(0) = 1 \ , \quad t,s \in R \ ,$$

(2) $$C(t+s) + C(t-s) = 2C(t)\,C(s) \ , \ C(0) = 1 \ , \quad t,s \in R \ .$$

It has been observed by A.Cauchy [6] that functional equations (1) and (2) are characteristic equations for functions E and C. This fact lies behind the idea to define exponential and cosine functions in abstract structures as functions which satisfy functional equations (1) and (2) and of course some kind of regularity conditions. The theory developed with such an idea for exponential function and differential equation $x' = Ax$ is well-known as a theory of semigroups [12]. The parallel theory for cosine function and differential equation $x'' + Ax = 0$ is known as the theory of abstract cosine functions, or cosine operator functions. Since the theory of semigroups is well-known we will be more concerned with abstract cosine functions.Questions concerning differential equation $x'' + Ax = 0$, cosine functions and infinitesimal generations will be treated in the paper by D.Lutz which follows our paper.

§ 1. SEMIGROUPS AND COSINE FUNCTIONS IN BANACH ALGEBRAS

1. In order to prove that exponential functions are characterized by functional equation

$$(1) \qquad E(t+s) = E(t)\,E(s)\ , \quad E(0) = 1\ , \qquad t,s \in R\ ,$$

suppose that $E: R \to C$ is a continuous function and that it satisfies (1). From this we find

$$(2) \qquad \lim_{t\to 0} \frac{1}{t} \int_0^t E(s)\,ds = E(0) = 1\ .$$

Hence there exists a number $\gamma \neq 0$ such that

$$(3) \qquad \Gamma = \int_0^\gamma E(s)\,ds \neq 0\ .$$

Multiplying (3) by $E(t)$ we get

$$(4) \qquad \Gamma E(t) = \int_0^\gamma E(s)\,E(t)\,ds = \int_0^\gamma E(t+s)\,ds = \int_t^{\gamma+t} E(s)\,ds$$

from which we find that E is differentiable. Taking derivative of (1) with respect to s and setting $E'(0) = a$ we get

$$E'(t) = a\,E(t) \Rightarrow E(t) = 1 + a\int_0^t E(s)\,ds$$

from which by iteration we get

$$(5) \qquad E(t) = 1 + \frac{at}{1!} + \frac{a^2t^2}{2!} + \ldots + \frac{a^nt^n}{n!} + \ldots\ , \quad t \in R\ .$$

2. Now we take a continuous function $t \to E(t)$ from the set R into the set of all $n \times n$ complex matrices and we assume that (1) holds. That $t \to E(t)$ is continuous means that all the functions $t \to (E(t))_{ij}$ $(i,j = 1,\ldots,n)$ are continuous on R. Since $E(0) = 1 = (\delta_{ij})$ we have

$$(6) \qquad \lim \frac{1}{t} \int_0^t (E(t))_{ij}\,dt = \delta_{ij}$$

from which it follows that matrices M_α :

$$(M_\alpha)_{ij} = \int_0^\alpha (E(t))_{ij}\, dt$$

have the property that $\frac{1}{\alpha} M_\alpha \to 1$. But then $\det(\frac{1}{\alpha} M_\alpha) = \alpha^{-n} \det M_\alpha \to 1$. Hence there is a number γ such that $\det M_\gamma \neq 0$. From (1) we get

$$(E(t+s))_{ij} = \sum_{p=1}^{n} (E(t))_{ip} (E(s))_{pj} ,$$

(7) $$\int_0^\alpha (E(t+s))_{ij}\, ds = \sum_{p=1}^{n} (E(t))_{ip} \int_0^\alpha (E(s))_{pj}\, ds .$$

For $\alpha = \gamma$ from (7) we get

(8) $$\sum_{p=1}^{n} (E(t))_{ip} C_{pj} = \int_t^{\gamma+t} (E(s))_{ij}\, ds$$

with $C = M_\gamma$.

Now multiply (8) by $(C^{-1})_{jk}$ and sum over j to get:

(9) $$(E(t))_{ik} = \sum_{j=1}^{n} C^{-1}_{jk} \int_\gamma^{\gamma+t} (E(s))_{ij}\, ds .$$

From (9) we conclude that all the functions $t \to (E(t))_{ik}$ are differentiable on R. By taking derivative of $E(t+s) = E(t)\,E(s) = E(s)\,E(t)$ with respect to s, for $s = 0$ we get

(10) $$E'(t) = AE(t) = E(t)A , \quad t \in R , \quad A = E'(0) .$$

From (10) it follows that

(11) $$E(t) = 1 + \frac{At}{1!} + \frac{A^2t^2}{2!} + \dots + \frac{A^nt^n}{n!} + \dots , \quad t \in R$$

which we write in the form

(12) $$E(t) = \exp t A , \quad t \in R .$$

3. Let A be a Banach algebra with an identity 1 and let $t \to E(t)$ be a continuous function from $R^+ = [0,\infty)$ into A such that

(13) $$E(t+s) = E(t)\, E(s),\ E(0) = 1 , \quad t,s \in R^+ .$$

A function E which satisfies (13) is called a <u>semigroup</u> in A. The continuity of $t \to E(t)$ in $t = 0$ implies

$$\lim_{t\to 0} \frac{1}{t} \int_0^t E(s)\, ds = 1$$

where the integral in A is taken in the Bochner sense. Hence there is a number $\gamma > 0$ such that

$$\left| 1 - \frac{1}{\gamma} \int_0^\gamma E(s)\, ds \right| < 1 ,$$

so that $\frac{1}{\gamma}\int_0^\gamma E(s)\, ds$ is a regular element of A . Thus $C = \int_0^\gamma E(s)\, ds$ is also a regular element of A . Integrating (13) from 0 to γ we get

$$\int_0^\gamma E(t+s)\, ds = E(t) \int_0^\gamma E(s) ds = CE(t) ,$$

$$E(t) = C^{-1} \int_t^{\gamma+t} E(s)\, ds .$$

Hence the function E is differentiable. But then

$$E'(t) = a\, E(t) , \quad a = E'(0)$$

implies

$$E(t) = 1 + a \int_0^t E(s)\, ds$$

which by the method of succesive approximations leads to

$$(14) \qquad E(t) = \sum_{n=0}^{\infty} \frac{a^n t^n}{n!} \qquad (t \in R^+) .$$

By power series $\sum_{n=0}^{\infty} \frac{a^n t^n}{n!}$ which converges for each $a \in A$ and every $t \in R$ we define the function $e^{ta} = \exp ta$. It is easy to see that this function is continuous and that it satisfies the functional equation (1). For some further results on semigroups in Banach algebras see [12].

4. If A is a Banach algebra with an identity 1 and $C: R \to A$ a function such that

$$(15) \qquad C(t+s) + C(t-s) = 2C(t)\, C(s) , \qquad t,s \in R$$

$$C(0) = 1$$

holds true then we say that C is a <u>cosine function</u> in A .

THEOREM 1.([29]). Let A be a Banach algebra with an identity 1 and $C:R\to A$ a function which satisfies (15).

If C is continuous, then there exists an element $a\in A$ such that

$$C(t) = 1 + \frac{at^2}{2!} + \ldots + \frac{a^n t^{2n}}{(2n)!} + \ldots , \qquad t\in R \quad . \tag{16}$$

PROOF. Continuity of C and $C(0) = 1$ implies

$$\lim \frac{1}{t} \int_0^t C(s)\, ds = 1 .$$

Hence there is a number γ such that

$$|1 - \frac{1}{\gamma} \int_0^\gamma C(s)\, ds | < 1 \quad \Rightarrow \quad \int_0^\gamma C(s)\, ds$$

is a regular element of A. If we integrate (15) from 0 to α with respect to s we get

$$2C(t) \int_0^\alpha C(s)\, ds = \int_t^{t+\alpha} C(s)\, ds + \int_{-t}^{-t+\alpha} C(s)\, ds ,$$

$$C(t) = \frac{1}{2} (\int_t^{t+\gamma} C(s)ds - \int_t^{t-\gamma} C(s)ds)\cdot(\int_0^\gamma C(s)\, ds)^{-1} . \tag{17}$$

From (17) we find the function C to be differentiable on R and

$$C'(t) = \frac{1}{2} (C(t+\gamma) - C(t-\gamma)), \quad C'(0) = 0 . \tag{18}$$

From (18) we conclude that C has derivative of any order. By taking second derivative of (15) with respect to s, for $s = 0$ we get

$$C''(t) = a\, C(t) , \quad t\in R \tag{19}$$

with $a = C''(0)$. From (19) we find

$$C'(t) = a \int_0^t f(u)\, du , \; C(t) = 1 + a \int_0^t (\int_0^s C(u)du)\, ds \Rightarrow$$

$$C(t) = 1 + a \int_0^t (t-s)\, C(s)\, ds . \tag{20}$$

The iteration method applied to the integral equation (2) leads to

$$C(t) = 1 + \frac{at^2}{2!} + \ldots + \frac{a^n t^{2n}}{(2n)!} + \frac{a^{n+1}}{(2n+1)!} \int_0^t (t-s)^{2n+1} C(s)ds$$

from which (16) follows. □

If in A there is an element b such that $b^2 = a$, then (16) can be written in the form

(21) $$C(t) = \frac{1}{2} [\exp tb + \exp(-tb)] = \cosh tb, \quad t \in R$$

where the hyperbolic cosine of an element of a Banach algebra is defined by the series.

Generally the square root of a does not exist in A and the solution of (15) is not of the form (21).

However, it is not too far from that form, because as in [29] we can imbed the Banach algebra A into another Banach algebra $\hat{A}$ which has the property that any $a \in A$ as an element in $\hat{A}$ has a square root in $\hat{A}$, i.e. there is at least one element $\hat{b} \in \hat{A}$ such that $\hat{b}^2 = \hat{a}$. The definition of such a Banach algebra $\hat{A}$ is very simple. It is sufficient to consider all 2×2 matrices $\hat{x}$, $\hat{y}$, the elements x_{ij}, y_{ij} $(i,j = 1,2)$ of which are elements of A and to define the usual matrix operations between such matrices. Introducing the norm in $\hat{A}$ by the formula

$$|\hat{x}| = \sum_{i,j=1}^{2} |x_{ij}|$$

one easily verifies that $\hat{A}$ is a Banach algebra. In the Banach algebra $\hat{A}$ we imbed (isomorphically but not isometrically) the Banach algebra A by the correspondence:

$$a \to \hat{a} = \begin{pmatrix} a & 0 \\ 0 & a \end{pmatrix}, \qquad a \in A .$$

Now, simple calculation shows that

$$\hat{a}^2 = \begin{pmatrix} a & 0 \\ 0 & a \end{pmatrix}, \quad \hat{a} = \begin{pmatrix} 1+\frac{a}{4} & 1-\frac{a}{4} \\ -(1-\frac{a}{4}) & -(1+\frac{a}{4}) \end{pmatrix},$$

for every $a \in A$, i.e. in $\hat{A}$ every element $a \in A$ has a square root.

The function

$$\hat{C}(t) = \begin{pmatrix} C(t) & 0 \\ 0 & C(t) \end{pmatrix}, \quad t \in R$$

satisfies all the conditions of Theorem 1 and

$$\hat{C}(t) = \frac{1}{2} (\exp t\hat{a} + \exp (-t\hat{b})) = \cosh t\hat{a}$$

holds for every $t \in R$.

In connection with a problem of imbedding an algebra A into a large algebra in a way that any element of A in the large algebra has a square root we have proved the following theorem.

THEOREM 2.([28]) Let Φ be the field of real or complex numbers and $A = \{a,b,\dots\}$ a Banach algebra over Φ with a unit element 1.

Then, there exists a Banach algebra $A' = \{T,S,\dots\}$ over Φ with a unit E and an imbedding $j: A \to A'$ such that

1. $j(\alpha a + \beta b) = \alpha j(a) + \beta j(b)$
2. $j(a b) = j(a) j(b)$
3. $|j(a)| = |a|$
4. $j(a)$ and a have the same spectrum.

Furthermore, if a is any element of A and n any natural number, then there exists at least one element $T \in A'$ such that

I. $T^n = j(a)$,

II. if $b \in A$ commutes with a, then T commutes with $j(b)$.

Now, we apply Theorem 2 to (16). By setting $\hat{C}(t) = j(C(t))$ we get

$$\hat{C}(t) = E + \frac{a't^2}{2!} + \dots + \frac{a'^n t^{2n}}{(2n)!} + \dots$$

with $a' = j(a)$. Since a' has a square root $\hat{a}$ in $\hat{A}$ we get

$$\hat{C}(t) = \frac{1}{2} (\exp t\hat{a} + \exp (-t\hat{a})) = \cosh t\hat{a}\ .$$

We can summarize our discusion in the following way. If the function C is continuous then it can be expressed by the exponential function possibly in a larger algebra. J.A.Baker and K.R.Davidson [3] have proved that the continuity assumption is essential one for such reduction of a function C which satisfies (2) to a function E which satisfies

(1).

Theorem 1 has been generalized by J.A.Baker [2] .He has proved the following theorem.

THEOREM 3. ([2]). Let A be a Banach algebra and let $C:(0,\infty)\to A$ be such that

$$C(t+s) + C(t-s) = 2C(t)\,C(s) \tag{22}$$

whenever $t > s > 0$. If

$$\lim_{t\to 0} C(t) = j$$

exists then $j^2 = j$ and there exist elements $a,b \in A$ such that $ja = aj = a, bj = b, jb = 0$ and

$$C(t) = \left(j + \frac{at^2}{2!} + \frac{a^2t^4}{4!} + \ldots\right) + b\left(tj + \frac{t^3a}{3!} + \frac{t^5a^2}{5!} + \ldots\right) \tag{23}$$

for all $t > 0$. Conversely, with such j, a and b , if C is defined by (23) for all $t \in R$, then C satisfies (22) for all $t,s \in R$.

Since the proofs of Theorems 3 and 2 are rather long they will not be given here.

§ 2. MEASURABILITY AND CONTINUITY

In this section by X we denote a Banach space, by X^* the dual space of X and by $L(X)$ the set of all linear and continuous mappings of X into X endowed with the usual structure of a Banach space. A mapping

$$C: R \to L(X)$$

is called a cosine operator function if

(1) $$C(t+s) + C(t-s) = 2C(t)\,C(s)$$

and $C(0) = I$ holds for all $t, s \in R$, where I is the identity operator.

THEOREM 4. ([25] and [26]) . Let C be a cosine operator function on a Banach space X . Suppose that:

a) there is an interval $\Delta \subset R$ such that the restriction of C on Δ is weakly measurable in the Lebesgue sense, and

b) X is a separable and reflexive Banach space.

Then C is a weakly continuous function on R .

Let us remark that the weak continuity resp. the weak measurability of a function C on $\Delta = [a,b]$ means that for each $x \in X$ and each $y^* \in X^*$ the function $t \to y^*(C(t)x)$ is continuous resp. measurable in the Lebesgue sense on the interval Δ .

The proof of Theorem 4 for X a Hilbert space was done in [25] and X a reflexive space in [26] and it depends on the following lemma.

LEMMA 1. ([25]) Let K be a linear Lebesgue measurable set such that $0 < m(K) < +\infty$. There exists a number $a > 0$ with the property that for every $t \in (-a,a)$ there are $s_1(t), s_2(t), s_3(t) \in K$ such that

$$s_1(t) = s_2(t) - \frac{t}{2} = s_3(t) - t \quad .$$

PROOF. Let u be the function defined on R by the equation $u(t) = m[K \cap (K-\frac{t}{2}) \cap (K-t)]$. If $t \to k(t)$ denotes the characteristic function of the set K then

$$|u(t) - u(0)| =$$

$$= |\int k(s)\,[k(s+\tfrac{t}{2})\,k(s+t) - k(s)\,k(s+t) + k(s)\,k(s+t) - k(s)]\,ds|$$

$$\leq \int |k(s+\tfrac{t}{2}) - k(s)|\,ds + \int |k(s+t) - k(s)|\,ds .$$

Since the right hand side tends to zero as $t \to 0$ we find the function u continuous in $t = 0$. Since $u(0) = m(K) > 0$, there exists a constant $a > 0$ such that $u(t) \neq 0$ for each $t \in (-a,a)$. But $u(t) \neq 0$ implies $K \cap (K-\frac{t}{2}) \cap (K-t) \neq \emptyset$. Hence for each $t \in (-a,a)$ there are $s_1(t)$, $s_2(t)$, $s_3(t) \in K$ such that $s_1(t) = s_2(t) - \frac{t}{2} = s_3(t) - t$ and hence Lemma 1 is proved.

PROOF of Theorem 4.

1. <u>The function C is measurable on R.</u> (1) implies:

$$C(t - \tfrac{b-a}{2}) = 2C(t)\,C(\tfrac{b-a}{2}) - C(t + \tfrac{b-a}{2}) .$$

When t runs through the interval $[a, \frac{1}{2}(a+b)]$ then $t + \frac{1}{2}(b-a)$ runs over the interval $[\frac{1}{2}(a+b), b]$. Since $s \to C(s)$ is measurable on each of these intervals we find that C is measurable on the interval $[a - \frac{1}{2}(b-a), a]$. Thus the measurability of C on the interval Δ implies the measurability of this function on the interval $\Delta' = [a - \frac{1}{2}(b-a), b]$. The way by which Δ' is obtained from Δ enables us to deduce that the function C is measurable on the set $(-\infty, b)$. For $t = 0$ (1) implies that C is an even function. Thus the function C is measurable on the set of all real numbers.

2. <u>The function C is locally bounded</u>. Since $X = (X^*)^*$ is separable the space X^* is also separable. If $x_1, x_2, \ldots$ is a dense set on the unit sphere of X and $y_2^*, y_2^*, \ldots$ a dense set on the unit sphere of X^* then

$$\|C(t)\| = \sup\{\,|y_i^*(C(t)x_j)| : i,j \in N\}$$

and measurability of functions $t \to y_i^*(C(t)x_j)$ imply that $t \to \|C(t)\|$ is a measurable function on R. Hence there is a measurable set $K \subset R$

of strictly positive measure such that

$$L = \sup \{ \|C(t)\| : t \in K \} < \infty .$$

We assert that $t \to \|C(t)\|$ is bounded on every finite interval. Since the function C is an even function we can without loss of generality, assume that $K \subseteq [0,+\infty)$. If we put $t+s$ instead of s in (1) we get

$$C(t) = 2C(t+s)\, C(s) - C(t+2s) .$$

This implies:

(2) $$\|C(t)\| \leq 2\| C(t+s)\| \cdot \| C(s)\| + \|C(t+2s)\| .$$

For $t = s$ (1) implies $C(2t) = 2C(t)^2 - I$ and this gives:

(3) $$\|C(2t)\| \leq 2 \|C(t)\|^2 + 1 .$$

From (2) and (3) we get :

(4) $$\|C(t)\| \leq 2 \|C(t+s)\| \cdot \| C(s)\| + 2 \|C(s+ \tfrac{1}{2} t)\|^2 + 1 .$$

According to Lemma 1 there exists a number $a > 0$ with the property that for every $t \in (0,a)$ a number s can be found such that $s, s + \frac{1}{2} t, s + t \in K$. If $t \in (0,a)$ and if s is the corresponding element of K then (4) implies: $\| C(t) \| \leq 4L^2+1$ for every $t \in (0,a)$. Thus the function $t \to \|C(t)\|$ is bounded on the interval $(0,a)$. This and (3) imply that $t \to \|C(t)\|$ is bounded on the interval $(0,2a)$. From this we find that the function $t \to \|C(t)\|$ is bounded on every finite interval of the type $(0,b)$, $b > 0$. Since C is an even function we have that it is bounded on every finite interval.

3. <u>The function C is weakly continuous</u>. Since the function C is measurable and locally bounded, the function

$$y^* [C(t)x] \qquad (x \in X , \; y^* \in X^*)$$

is summable on every finite interval. This implies that the equation

(5) $$y^*_{ab}(x) = \int_a^b y^* [C(t)x]\, dt$$

defines a bounded linear functional y^*_{ab} for any $a,b \in R$ and $y^* \in X^*$. By X^*_1 we denote the set of all functionals $y^*_{ab} \in X^*$ which can be written in the form (5). We assert that X^*_1 is dense in X^*. Suppose that this is not true. Then a functional $z^* \in X^*$ exists such that $z^* \notin \overline{X}^*_1$ and $z^* \neq 0$. But then a functional $w^{**} \in X^{**}$ can be found such that

(6) $$w^{**}(z^*) = 1 , \quad w^{**}(y^*) = 0 , \quad y^* \in X^*_1 .$$

Since X is a reflexive space one can find an element $w \in X$ such that

$$w^{**}(y^*) = y^*(w) , \quad y^* \in X^* .$$

This and (6) lead to

(7) $$y^*_{ab}(w) = 0$$

for all $y^*_{ab} \in X^*_1$. Now (7) and (5) imply

(8) $$\int_a^b y^*[C(t)w]\,dt = 0$$

for every couple $a,b \in R$ and for any $y^* \in X^*$. From (8) we find

(9) $$y^*(C(t)w) = 0$$

for every $t \notin S(y^*)$ where $mS(y^*) = 0$ (the Lebesgue measure of $S(y^*)$). Since X^* is a separable space there is a set $A = \{y^*_1, y^*_2, \ldots, y^*_n, \ldots\}$ which is countable and dense on X^*. We set $S = \bigcup_{k=1}^{\infty} S(y^*_k)$. Then $mS = 0$ and $y^*_n(C(t)w) = 0$ for every $t \notin S$. Since A is dense on X^* we find

(10) $$C(t)w = 0$$

for every $t \notin S$. Obviously $mS = 0$ implies the existence of a number $u \in R$ such that $u \notin S$ and $2u \notin S$. From (10) we find $C(2u)w = C(u)w = 0$, which together with the functional equation (1) leads to $w = C(0)w = [C(2u) - 2C(u)^2]\, w = 0$, which contradicts (6). Thus X^*_1 is dense on X^*. Replacing t by $2C(t)x$ in (5) and using (4) we get

$$2y^*_{ab}[C(s)x] = \int_a^b y^* [2C(s)\ C(t)x]\ dt$$

$$= \int_a^b y^* [(C(t+s)+C(t-s)x]\ dt = \int_{a+s}^{b+s} y^* [C(t)x]dt + \int_{a-s}^{b-s} y^*[C(t)x]\ dt\ ,$$

from which we see that the function $t \to z^*[C(t)x]$ is continuous on R for any $z^* \in X^*_1$. Since X^*_1 is dense on X^* and $t \to \|C(t)\|$ is locally bounded, we find that $t \to y^*[C(t)x]$ is continuous function on R for any pair $x \in X$ and $y^* \in X^*$, i.e. C is a weakly continuous function on R . □

THEOREM 5. ([2]) . Let G be a locally compact Abelian group, let A be a Banach algebra and suppose $C:G \to A$ satisfies (1) for all $t,s \in G$. If C is strongly measurable on a set of positive, finite Haar measure, then the mapping $t \to C(2t)$ is continuous. If in addition the mapping $t \to 2t$ is a bicontinuous automorphism of G , then C is continuous at 0 .

PROOF.Suppose C is strongly measurable on a measurable set S of positive finite Haar measure. Then, by the definition of strong measurability, C is the pointwise limit almost everywhere on S of a sequence of countably valued measurable functions. As in the complex valued case, the theorems of Egorov and Lusin can be proved and we conclude that there exists a compact subset K of S of positive Haar measure such that the restriction of C to K is continuous. It follows that C is also uniformly continuous on K .

Since K has a positive finite Haar measure, as in Lemma 1 , there exists a neighbourhood V of $0 \in G$ such that

$$K \cap (K+v) \cap (K-v) \neq \emptyset$$

for each $v \in V$.

Let $\varepsilon > 0$ and $L = \max\{\|C(t)\| : t \in K\}$. Since C is uniformly continuous on K there exists a symmetric neighbourhood U of $0 \in G$ such that $\|C(s)-C(t)\| < \varepsilon/4L$ provided $s,t \in K$ and $s-t \in U$. Now (1) for $t+s = 2u$, $t-1 = 2v$ implies

$$C(2v) + C(2u) = 2C(u+v)\ C(u-v)$$

and for $t = s = v$

$$C(2u) + C(0) = 2C(u)\, C(u) .$$

From here we get

$$\|C(2v) - C(0)\| = 2\,\|C(u+v)\, C(u-v) - C(u)\, C(u)\,\|$$

$$\leq 2\,\|C(u+v)\| \cdot \|C(u-v) - C(u)\| + 2\,\|C(u)\| \cdot \|C(u+v) - C(u)\| .$$

If $v \in V \cap U$ then there exists $u \in K$ such that $u+v \in K$ and $u-v \in K$ so that $v \in V \cap U$ implies $\|C(2v) - C(0)\| < \varepsilon$. □

THEOREM 6.([12]). Let X be a Banach space and $E:(0,\infty) \to L(X)$ a semigroup in $L(X)$, i.e.

$$(11) \qquad E(t+s) = E(t)\, E(s) \qquad t,s \in (0,\infty) .$$

If a function E is strongly measurable, then it is strongly continuous.

LEMMA 2. ([12]) Let $E:(0,\infty) \to L(X)$ be a semigroup and $x \in X$. If a function $t \to E(t)x$ is strongly measurable on $(0,\infty)$, then it is locally bounded.

PROOF. Suppose that the function $t \to E(t)x$ is not bounded on some interval $[a,b]$, $a > 0$. Then a number $\tau_o \in [a,b]$ and a sequence (τ_n) in $[a,b]$ exist such that

$$\|E(\tau_n)x\| > n \quad , \qquad \tau_n \to \tau_o .$$

Since $t \to E(t)x$ is strongly measurable the numerical function $t \to \|E(t)x\|$ is also measurable. Hence there exists a measurable set $K \subset [0,\tau_o]$ such that $m(K) > \frac{\tau_o}{2}$ and the function $t \to \|E(t)x\|$ is bounded on K, i.e. $\|E(t)x\| \leq M$ for all $t \in K$ with some real number $M > 0$. Set

$$S_n = \{\, \tau_n - s : s \in K \cap [0,\tau_n] \} .$$

The sets S_n are measurable and $m(S_n) \geq \gamma/2$. If $s \in K \cap [0,\tau_n]$ then

$$n \leq \|E(\tau_n)x| \leq \|E(\tau_n - s)\| \cdot \|E(s)x\| \leq M\,\|E(\tau_n - s)\| .$$

Hence $\|E(t)\| \geq n/M$ for all $t \in S_n$. Denoting $\limsup S_n$ by S we find $m(S) \geq \tau_o/2$ and $\|E(t)\| = +\infty$ for each $t \in S$. This contradicts the assumption that $E(t)$ is a bounded operator for $t > 0$.

PROOF of Theorem 6. Since the function $t \to E(t)x$ is strongly measurable it is locally bounded for each $x \in X$, i.e. for any interval $[a,b]$, $a > 0$ we have

$$(12) \qquad \sup\{\|E(t)x\| : t \in [a,b]\} < +\infty, \quad x \in X.$$

By the principle of uniform boundedness (12) implies

$$M = \sup\{\|E(t)\| : t \in [a,b]\} < +\infty,$$

i.e. the function $t \to \|E(t)\|$ is locally bounded. Hence the function $t \to E(t)x$ is locally integrable for each $x \in X$. If we set $u = t+s$ in (11) we get

$$E(u)x = E(t)\,E(u-t)x \quad \Rightarrow$$

$$(b-a)\;E(u)x = \int_a^b E(t)\,E(u-t)x\,dt \quad \Rightarrow$$

$$(b-a)\;\|E(u)x - E(u_0)x\| = \left\|\int_a^b E(t)\left[E(u-t)x - E(u_0-t)x\right]dt\right\|$$

$$\leq M \int_a^b \|E(u-t)x - E(u_0-t)x\|\,dt.$$

But the right hand side of the last inequality tends to zero as $u \to u_o$. Thus

$$\|E(u)x - E(u_o)x\| \to 0 \qquad \text{as } u \to u_o.$$

THEOREM 7. Let X be a Banach space and $C:(0,\infty) \to L(X)$ a function such that (1) holds for all $0 < s \leq t$; $s,t \in (0,\infty)$. If the function $t \to C(t)$ is strongly measurable then it is strongly continuous.

LEMMA 3. ([5]). Let $C:(0,\infty) \to L(X)$ be a function such that (1) holds for all $0 < s \leq t$; $s,t \in (0,\infty)$. If for some $x \in X$ the function $t \to C(t)x$ is strongly measurable, then it is locally bounded.

PROOF. The function $t \to \|C(t)x\|$ is measurable. Suppose, that there exists an interval $[a,b]$, $a > 0$ such that the function $t \to \|C(t)x\|$

is not bounded on $[a,b]$. Then a sequence $\tau_o, \tau_1, \tau_2, \ldots$ exists such that

(12) $$\tau_n \downarrow \tau_o, \quad \| C(\tau_n)x \| \geq n \qquad (n = 1,2,3,\ldots) .$$

Since $t \to \| C(t)x \|$ is measurable there exists a constant $M > 0$ and a measurable set $G \subset [0, \tau_o]$ such that

(13) $$m(G) > \frac{3}{4} \tau_o , \quad \| C(t)x \| \leq M , \ t \in G .$$

Set

(14) $$A_k = \frac{1}{2} \tau_k - \frac{1}{2} (G \cap [0, \tau_k]), B_k = G \cap [0, \frac{1}{2} \tau_k] \ (k=0,1,2,\ldots)$$

We claim that $m(A_o \cap B_o) > 0$. Otherwise we would have $m(A_o \cap B_o) = 0$ so that $mA_o + mB_o \leq \frac{1}{2} \tau_o$. But $mA_o = \frac{1}{2} m G$. From $2mA_o + 2mB_o \leq \tau_o$ we get $\frac{3}{4} \tau_o < m G \leq \tau_o - 2mB_o$, i.e. $m(B_o) < \frac{1}{8} \tau_o$. Also

$$G = (G \cap [0, \frac{1}{2} \tau_o]) \cup (G \cap [\frac{1}{2} \tau_o, \tau_o]) = B_o \cup C_o ,$$

with $C_o = G \cap [\frac{1}{2} \tau_o, \tau_o]$. Therefore

$$mG = MB_o + mC_o \quad \text{and} \quad mC_o \leq \frac{1}{2} \tau_o ,$$

so that

$$\frac{3}{4} \tau_o < mG = mB_o + mC_o \leq mB_o + \frac{1}{2} \tau_o ,$$

i.e.

$$mB_o > \frac{1}{4} \tau_o .$$

Since $mB_o > \frac{1}{4} \tau_o$ and $mB_o < \frac{1}{8} \tau_o$ contradict each other we have $m(A_o \cap B_o) > 0$. There exists a number $\delta > 0$ such that

$$m(A_o \cap B_o) > \delta .$$

Set

$$E_k = A_k \cap B_k , \qquad (k = 0,1,2,\ldots)$$

and

$$H_n = \tau_n - E_n \qquad (n=1,2,\ldots)\ .$$

Now $\tau_n \to \tau_o$ imples $E_n \to E_o$. Therefore for sufficiently large $n, m(E_n) > 0$ and for such values of n, if $s \in E_n$ one can easily see that s and $\tau_n - 2s$ both belong to G . Obviously for sufficiently large $n, m(H_n) \geq \frac{1}{2}\delta$. For $s \in E_n$ by using equation (1) we get:

$$n \leq \|C(\tau_n)x\| = \|C[(\tau_n - s)+s]x\| = \|2C(\tau_n - s)\ C(s)x - C(\tau_n - 2s)x\|$$

$$\leq 2\|C(\tau_n - s)\| \cdot \|C(s)x\| + \|C(\tau_n - 2s)x\| \leq 2M\|C(\tau_n - s)\| + M\ .$$

Hence

$$\frac{n-M}{2M} \leq \|C(\tau_n - s)\|\ , \qquad s \in E_n\ ,$$

i.e.

(15) $$\frac{n-M}{2M} \leq \|C(t)\|\ , \quad t \in H_n\ .$$

Denoting lim sup H_n by H , we see that $m(H) \geq \frac{1}{2}\delta$ and $\|C(t)\| = +\infty$ for each $t \in H$ which contradicts the assumption that $C(t) \in L(X)$. □

PROOF of Theorem 7. Lemma 3 implies

$$M = \sup\{\|C(t)\| : t \in [a,b]\} < \infty$$

for any interval $[a,b]$, $a > 0$. For $x \in X$ we have

$$\left.\begin{array}{l} C(u) = 2C(s)\ C(u-s) - C(u-2s) \\ C(u_o) = 2C(s)\ C(u_o - s) - C(u_o - 2s) \end{array}\right\} \Rightarrow$$

$$C(u)x - C(u_o)x = 2C(s)\,[C(u-s) - C(u_o - 2s)]\,x$$

$$- [C(u-2s) - C(u_o - 2s)]\,x \qquad \Rightarrow$$

$$(b-a)\,[C(u)x - C(u_o)x] = 2\int_a^b C(s)\,[C(u-s) - C(u_o - 2s)]\,x\,ds$$

$$- \int_a^b [C(u-2s) - C(u_o - 2s)]\,x\,ds \quad \Rightarrow$$

$$(b-a)\| C(u)x - C(u_o)x \| \leq 2M \int_a^b \|C(u-s)x - C(u_o-2s)x\| \, ds$$

$$+ \int_a^b \| C(u-2s)x - C(u_o-2s)x \| \, ds$$

which tends to zero as $u \to u_o$. Thus

$$\| C(u)x - C(u_o)x \| \to 0 \qquad \text{as } u \to u_o \ . \ \square$$

§3. SPECTRAL REPRESENTATIONS FOR COSINE OPERATOR FUNCTIONS

One of the most important theorems in representation theory is the theorem obtained in 1932 by M.H.Stone:

If $t \to U(t)$ is a weakly continuous unitary representation of the additive group of real numbers in a Hilbert space X, then

$$(1) \qquad U(t) = \exp(tiA), \qquad t \in R$$

where A is a selfadjoint operator. Since A may be unbounded the exponential function in (1) is defined by use the spectral theorem for selfadjoint operators:

$$U(t) = \int_R e^{it\lambda} E(\Delta_\lambda)$$

where $E(\Delta)$ is the spectral measure associated with the operator A.

Similar result holds for a semigroup of normal operators: If $t \to N(t)$ is a weakly continuous semigroup of bounded normal operators on a Hilbert space X, i.e.

$$N(t+s) = N(t)\,N(s) \qquad (t,s \in [0,\infty))$$

and if $N(t)x = 0$ implies $k = 0$ for each t, then

$$(2) \qquad N(t) = \exp tA, \qquad t \in [0,\infty)$$

where A is a normal in general unbounded operator [42].

Representations by unbounded selfadjoint operators were considered by: A.Devinatz [7] in 1954, S.Kurepa [18] in 1954, by A.Nussbaum [39] in 1959 and others. In this section we present an analogous result to (1) or (2) for a cosine operator function which takes values in the set of all bounded normal operators in a Hilbert space.

THEOREM 8. ([25]). Let X be a Hilbert space and $N: R \to L(X)$ a mapping such that

$$(3) \qquad N(t+s) + N(t-s) = 2N(t)\,N(s)$$

holds for all $t,s \in R$.

Suppose that: (1) $N(t)$ is a normal operator for every $t \in R$; (2) if $N(t)x = 0$, almost everywhere, then $x = 0$; (3) $t \to N(t)$ is a weakly continuous function.

Then a bounded selfadjoint operator B and a selfadjoint operator A which commutes with B can be found in such a way that

$$N(t) = \frac{1}{2}\left[\exp(itN) + \exp(-itN)\right] = \cos tN$$

holds for all $t \in R$ where $N = A + iB$.

LEMMA 4. ([21]). Let G be the set of all real numbers r of the form $m/2^n$ where m is an integer and n a natural number.

Suppose that $\phi : G \to C$ is such that

$$\phi(r+r') + \phi(r'-r) = 2\,\phi(r)\,\phi(r') \tag{4}$$

holds for all $r,r' \in G$. If there exists a natural number n_o such that

$$k \geq n_o \Rightarrow \operatorname{Re}\phi\left(\frac{1}{2^k}\right) > 0 , \tag{5}$$

then

$$\phi(r) = \cos ar , \quad r \in G$$

with a complex number a.

PROOF. From (5) it follows that $\phi \neq 0$. For $r' = r = 0$ (4) implies $\phi(0) = 1$. Hence the number $\phi(2^{-n_o})$ can be written in the form

$$\phi\left(\frac{1}{2^{n_o}}\right) = \frac{1}{2}\left[\exp(\alpha \pm i\beta) + \exp[-(\alpha \pm i\beta)]\right]$$

where α is a real number and $\beta \in [0, \pi/2]$. However only one sign is to be taken. For $r' = r$ (4) and $\phi(0) = 1$ imply:

$$\phi^2(r) = \frac{1+\phi(2r)}{2} .$$

Since $\operatorname{Re}\phi(1/2^{n_o+1}) > 0$ we have

$$\phi\left(\frac{1}{2^{n_o+1}}\right) = \frac{1}{2}\left[\exp\frac{\alpha \pm i\beta}{2} + \exp\left(-\frac{\alpha \pm i\beta}{2}\right)\right] .$$

In the same way we get:

$$\phi\left(\frac{1}{2^{n_o+k}}\right) = \frac{\rho+\rho^{-1}}{2}, \quad \rho = \exp \frac{\alpha \pm i\beta}{2^k} .$$

Now (4) implies $\phi(2r) = 2\,\phi(r)^2 - 1$ so that

$$\phi\left(\frac{2}{2^{n_o+k}}\right) = \frac{\rho^2+\rho^{-2}}{2}$$

and by induction we find :

$$\phi\left(\frac{m}{2^{n_o+k}}\right) = \frac{1}{2}\left[\exp \frac{m}{2^k}(\alpha \pm i\beta) + \exp\left(-\frac{m}{2^k}(\alpha \pm i\beta)\right)\right]$$

from which follows $\phi(r) = \cos ar$ for all $r > 0$, $r \in G$. Since ϕ is an even function and $\phi(0) = 1$ we find $\phi(r) = \cos ar$ for each $r \in G$.

PROOF of Theorem 8. I. For $y \in X$ the equation

$$(6) \qquad (y_{ab}|x) = \int_a^b (N(t)x|y)\,dt$$

defines a vector $y_{ab} \in X$. We assert that the set X' of all y_{ab} is dense in X . In fact if $z \in X$ is orthogonal to X' , then

$$(7) \qquad \int_a^b (N(t)z|y)\,dt = 0$$

for all $a,b \in R$ and $y \in X$. The continuity of the function $t \to (N(t)z|y)$ with (7) imply

$$(8) \qquad (N(t)z|y) = 0$$

for all $t \in R$ and for every $y \in X$. From here we get $N(t) = 0$ for all t which implies $z = 0$. Thus the set X' is dense in X . Using (3) we obtain:

$$\left(\frac{N(t)-I}{t}x\,\Big|\,y_{ab}\right) = \frac{1}{2t}\left[\int_b^{b+t}(N(u)x|y)\,du + \int_b^{b-t}(N(u)x|y)\,du - \int_a^{a+t}(N(u)x|y)\,du - \int_a^{a-t}(N(u)x\ y)\,du\right]$$

which implies

$$\lim_{t\to 0}\left(\frac{N(t)-I}{t}x\,\Big|\,y_{ab}\right) = 0$$

for every $y_{ab} \in X'$ and for every $x \in X$. From here it follows that

$$\frac{N^*(t)-I}{t} z$$

converges weakly to zero for every $z \in X'$, when $t \to 0$. There exists, therefore, a real number $M(z)$ such that:

$$\| (N(2^{-n})-I)z \| \leq 2^{-n} M(z) .$$

This implies that the series

$$(9) \qquad \sum_{n=1}^{\infty} \| (N(2^{-n})-I)z \|^2$$

is convergent for every $z \in X$!

II. Now we consider the restriction of the function $t \to N(t)$ to the set $G = \{\frac{k}{2^m}: m \in N , k \in Z\}$. Since $N(r)$ and $N(r')$ commute and since G is countable we find

$$(10) \qquad N(r) = \int_R f(\xi, r)\, E(\Delta_\xi)$$

where $E(\Delta)$ is a real spectral measure and the function $\xi \to f(\xi, r)$ is $E(\Delta)$-measurable and finite everywhere for every $r \in G$. If we put (10) in (3) we get

$$(11) \qquad f(\xi, r+r') + f(\xi, r'-r) = 2f(\xi, r)\, f(\xi, r')$$

for all $r, r' \in G$ and for almost all ξ (G is countable!). Using (10) we can write (9) in the form

$$(12) \qquad \lim_{n\to\infty} \int_R \sum_{k=1}^{n} |f(\xi, 2^{-n}) - 1|^2 \| E(\Delta_\xi) z \|^2 .$$

From (12) it follows that the series

$$(13) \qquad \sum_{n=1}^{\infty} |f(\xi, 2^{-n}) - 1|^2$$

is convergent almost everywhere with respect to the measure $\| E(\Delta) z \|^2$. Since the set X' is dense in X the series (13) is convergent almost everywhere with respect to $E(\Delta)$. Thus

(14) $$f(\xi, 2^{-n}) \to 1$$

almost everywhere with respect to $E(\Delta)$. It follows from (14),(11) and from Lemma 4 that

(15) $$f(\xi, r) = \frac{1}{2}[\exp ir\phi(\xi) + \exp(-ir\phi(\xi))]$$

holds true almost everywhere in ξ and for all $r \in G$. Here ϕ is an $E(\Delta)$-measurable and everywhere finite complex valued function. Thus the operators

(16) $$N = \int_R \phi(\xi)\, E(\Delta_\xi),\ A = \int_R [\mathrm{Re}\, \phi(\xi)]\, E(\Delta_\xi),\ B = \int_R [\mathrm{Im}\, \phi(\xi)] E(\Delta_\xi)$$

are defined. Since

$$\|N(r)\| = \mathrm{ess\ sup}\, |f(\xi, r)| < +\infty$$

for every $r \in G$, we find

$$\mathrm{ess\ sup} |\mathrm{Im}\, \phi(\xi)| < +\infty,$$

i.e. the operator B is bounded. Then (16), (15) and (10) imply:

$$N(r) = \frac{1}{2}[\exp(irN) + \exp(-irN)] = \cos rN$$

for every $r \in G$. By the weak continuity and the fact that the set G is dense on R we find: $N(t) = \cos(tN)$ for every $t \in R$. □

In the same way one can obtain representation (2) for a weakly continuous semigroup of normal operators. Theorem 8 has been generalized to cosine type representations of a locally compact Abelian groups by G.Maltese.

REFERENCES

[1] J.Aczél, Lectures on functional equations and their applications, New York, 1966.

[2] J.A.Baker, D'Alembert's functional equation in Banach algebras, Acta Sci. Math. Szeged, 32(1971), 225-234.

[3] J.A.Baker and K.R.Davidson, Cosine, exponential and quadratic functions, Glasnik Mat. 16(36)(1981), 269-274.

[4] A.B.Buche, On the cosine-sine operator functional equations, Aequ.Math. 6 (1971), 231-234.

[5] R.Chandler and H.Singh, On the measurability and continuity properties of the cosine operator, Indian J.pure appl. Math. 12(1)(1981), 81-83.

[6] A.Cauchy, Oeuvres complètes, IIe series, T.III, Paris (1897), 98-103 and 220-229.

[7] A.Devinatz, A note on semigroups of unbounded selfadjoint operators, Proc.Amer. Math.Soc.5(a954), 101-102.

[8] H.O.Fattorini, Uniformly bounded cosine functions in Hilbert spaces, Indiana Univ. Math. J. 20(1970),411-425.

[9] G.Faulkner and R.W.Shonkwiler, Cosine representations of Abelian *-semigroups and generalized cosine operator functions, Can.J.Math.,30(1978), 474-482.

[10] E.Giusti, Funzioni coseno periodiche, Boll.Unione Mat.Ital.22 (1967), 478-485.

[11] J.A.Goldstein, On the convergence and approximation of cosine functions, Aequ.Math. 10(1974), 201-205.

[12] E.Hille and R.S.Phillips, Functional analysis and semigroups, Amer.Math.Soc. Providence, R.I. 1957.

[13] Pl.Kannappan, The functional equation $f(xy)+f(xy^{-1}) = 2f(x)f(y)$ for groups, Proc. Amer. Math. Soc. 19(a968), 69-74.

[14] J.Kisynski, On operator-valued solutions of d'Alembert's functional equation I, Coll. Math. 23(1971), 107-114.

[15] J.Kisynski, On operator-valued solutions of d'Alembert's functional equation II, Studia Math. 24(a972),43-66.

[16] J.Kisynski, On cosine operator functions and one-parameter groups of operators, Studia Math. 44(1972), 93-105.

[17] H.Kraljević and S.Kurepa, Semigroups on Banach spaces, Glasnik Mat. 5(25)(1970), 109-117.

[18] S.Kurepa, Semigroups on unbounded self-adjoint transformations in Hilbert space, Glasnik mat.fiz. astr. 10(1955), 233-238.

[19] S.Kurepa, Convex functions, Glasnik mat.fiz. astr. 11(1956),89-94.

[20] S.Kurepa, Semigroups of linear transformations in n-dimensional vector space, Glasnik mat.fiz. astr. 13(1958),3-32.

[21] S.Kurepa, A cosine functional equation in n-dimensional vector space, Glasnik mat. fiz. astr. 13(1958), 168-189.

[22] S.Kurepa, On the continuity of semigroups of normal transformations in Hilbert space, Glasnik mat. fiz. astr. 13 (1958), 81-87.

[23] S.Kurepa, Semigroups of normal transformations in Hilbert space, Glasnik mat. fiz. astr. 13(1958), 257-266.

[24] S.Kurepa, On the quadratic functional, Publ. de l'Inst. Math.13 (1959), 57-75.

[25] S.Kurepa, A cosine functional equation in Hilbert space, Can.J. Math. 12(1960), 45-50.

[26] S.Kurepa, On some functional equations in Banach spaces, Studia Math. 19(1960), 149-158.

[27] S.Kurepa, A property of a set of positive measure and its applications, Journal Math. Soc. of Japan, 13(1961),13-19.

[28] S.Kurepa, On roots of an element of a Banach algebra, Publ. de l'Inst. Math. 1(11)(1962), 5-10.

[29] S.Kurepa, A cosine functional equation in Banach algebras, Acta Sci. Math. Szeged, 23(1962), 255-267.

[30] S.Kurepa, Uniformly bounded cosine function in Banach space, Mathematica Balkanica 2 (1972), 109-115.

[31] S.Kurepa, Weakly measurable selfadjoint cosine function, Glasnik Mat. 8 (28)(1973), 73-79.

[32] S.Kurepa, Decomposition of weakly measurable semigroups and cosine operator functions, Glasnik Mat. 11(31)(1976), 91-95.

[33] D.Lutz, Compactness properties of operator cosine functions, C.R.Math.Rep. Acad.Sci.Canada 2 (1980),277-280.

[34] D.Lutz, Über operatorwertige Lösungen der Funktionalgleichung des Cosinus, Math.7. 171(1980), 233-245.

[35] G.Maltese, Spectral representations for solutions of certain abstract functional equation, Compositio Math. 15 (1962),1-22.

[36] G.Maltese, Spectral representations for some unbounded normal operators, Trans.Amer.Math.Soc. 110(1964), 79-87.

[37] B.Nagy, On cosine operator functions on Banach spaces, Acta Sci. Math.Szeged 36(1974), 281-190.

[38] B.Nagy, Cosine operator functions and the abstract Cauchy problem, Period. Math. Hung. 7(1976), 213-217.

[39] A.E. Nussbaum, Integral representation of semigroups of unbounded self-adjoint operators, Ann.of Math.(2) 69(1959), 133-141.

[40] N.Sarapa, A note on the cosine equation for probability on compact semigroups, Glasnik Mat. 15(1980), 383-385.

[41] M.Sova, Cosine operator functions, Rosprawy Mat. XLIX(1966),3-46.

[42] B.Sz.Nagy, Spektraldarstellung linearer Transformation des Hilbertschen Raumes,Berlin, 1942.

[43] C.C.Travis and F.F.Webb, Compactness, regularity and uniform continuity properties of strongly continuous families, Houston J. Math. 3(1977), 555-567.

[44] F.Vajzović, Einige Funktionalgeichungen im Fréchetschen Raum, Glasnik Mat. 3(1968), 19-38.

STRONGLY CONTINUOUS OPERATOR COSINE FUNCTIONS

Dieter Lutz

1. INTRODUCTION

It was in 1821 that Cauchy gave a proof of the fact that the only continuous solutions $C:\mathbb{R}\to\mathbb{R}$ of d'Alembert's functional equation

$$(0) \qquad C(t+s) + C(t-s) = 2C(t)\,C(s), \quad s,t \in \mathbb{R}, \qquad C(0) = 1,$$

have the form $C(t) = \cos at$, $C(t) = \cosh at$ (with $a \in \mathbb{R}$) or $C(t) \equiv 1$ depending upon whether C is bounded and nonconstant, unbounded or constant on $\mathbb{R}$. The problem of proving analogous representation theorems under weaker assumptions, the discussion of related functional equations (especially those defining the pair of functions sine and cosine), and the study of connections of solutions of (0) and of Cauchy's functional equation of the exponential have since then constituted an important part of the theory of scalar-valued functional equations, as can be seen e.g. from Aczel's excellent monograph.

Compared with this well established theory, the theory of operator valued cosine functions attracted only fairly little attention. This is the more surprising as the theory of operator semi-groups was studied very closely by many authors during the last 30 years following the pioneering work by Hille, Phillips, Yoshida et al. An extraordinary amount of detailled informations concerning these systems of operators was given which led (among other applications) to a structural theory of evolution equations. One reason for the fact that cosine functions (which can be applied directly to linear second order problems) were neglected was that in many cases second order differential equations can be reduced to first order problems what makes semi-group theory applicable. It was not until Sova's work published in 1966 [39] that the basic facts on operator cosine functions especially with respect to the notion of an infinitesimal generator became known. Fattorini's papers from 1968/69 [5] did some parallel work but treated in more detail the connections with well-posed linear second order Cauchy problems for Banach space valued differential equations. A quite typical situation can be found already in these papers: Many results on operator semi-groups have natural analogues for cosine functions but the proofs in the latter case are more involved mostly due to the complicated structure of the functional equation under discussion. The problem to represent solutions C of (0) in the operator valued case as the cosine of an operator, problems of joint transformation to a selfadjoint or normal cosine function (both under additional assumptions as continuity of C in the operator norm topology, uniform boundedness on R a.s.o.) were tackled successfully by Kurepa and Fattorini. Nagy carried over several important results on spectral theory from semi-groups. Recently, interest focused on the connections with second order Cauchy problems. The practical impossibility to apply the Hille-Yoshida-type criterion for generators as given by Sova lead to the study of perturbation methods.

It is not the aim of this introduction to give credit to all people who contributed to cosine function theory. Nor is it the aim of the following four hours lecture to treat all the important achievements in this theory. We concernate on the task to give an introduction to some main problems and methods and stress the connection with problems in differential equations. All the given facts are well-known. Only in the first chapter we try to give a somewhat more straightforward treatment of the concept of infinitesimal generator. Proofs are only partly given in full length. More complete information is to be obtained from the papers listed in the bibliography.

2. BASIC FACTS ON OPERATOR COSINE FUNCTIONS AND THEIR GENERATORS

The facts treated in this chapter are mainly due to Sova [39], though we shall present the concept of infinitesimal generator in a somewhat different manner.

Let X always denote a complex Banach space and let $B(X)$ be the algebra of bounded linear operators on X.

2.1. DEFINITION. A function $C:R\to B(X)$ is called a strongly continuous operator cosine function on X if and only if

(i) $C(t+s) + C(t-s) = 2C(t)\,C(s), \quad s,t\in R$,

(1) (ii) $C(0) = I$,

(iii) $t\to C(t)x$ is continuous on R for each fixed $x\in X$.

2.2. EXAMPLES.

1. $X = \mathbb{C}$, $a\in\mathbb{C}$, $C(t) := \cosh\sqrt{a}t := \sum_{n=0}^{\infty} \frac{t^{2n}}{(2n)!}\, a^n$.

2. $X = R^2$, $C(t) := \begin{pmatrix} 1 & 0 \\ \frac{1}{2}t^2 & 1 \end{pmatrix}$, $t\in R$.

3. $X = C_o(R) := \{f:R\to\mathbb{C} \mid f \text{ continuous and bounded}\}$,

$\|f\| := \sup_{t\in R} |f(t)|$

$(C(t)f)(s) := \frac{1}{2}(f(s+t) + f(s-t)), \quad s,t\in R$.

The first two of these examples represent uniformly continuous operator cosine functions C that is $C:R\to B(X)$ fulfilling 2.1(i), (ii) and

(iii)′ $t\to C(t)$ is continuous on R with respect to the operator norm topology on $B(X)$.

Directly from 2.1. we get that C is even

$$C(s) + C(-s) = 2C(s)$$

(2)

$$C(s) = C(-s), \quad s\in R,$$

and that the values of C commute

$$2C(t)\,C(s) = C(t+s) + C(t-s)$$
$$= C(s+t) + C(s-t)$$
$$= 2C(s)\,C(t)\ ,\quad s,t \in R\ .$$

2.3. THEOREM. If C is a strongly continuous operator cosine function then there are constants $M \geq 1$ and $\omega \geq 0$ such that

$$\|C(t)\| \leq M\,e^{\omega|t|} \tag{3}$$

for all $t \in R$.

PROOF. Obviously we need to prove (3) only for $t \geq 0$. By Banach-Steinhaus theorem there is $K \geq 0$ with

$$\|C(t)\| \leq K \quad \text{for} \quad t \in [0,1]\ ,$$

and from 2.1.(ii) it follows that $K \geq 1$. Using d'Alembert's functional equation 2.1.(i) it is easily shown by induction that

$$\|C(nt)\| \leq (3K)^n$$

for all $t \in [0,1]$, $n \in N$.
Now take $s \in R^+$. There is exactly one $n \in N$ with $n-1 < s \leq n$. Putting $t := \frac{s}{n}$ we get

$$\|C(s)\| = \|C(tn)\| \leq (3K)^n = (3K)^s\,(3K)^{n-s} \leq (3K)^s\,3K$$

since $3K \geq 1$, $n-s \leq 1$. Put $M := 3K$ and $\omega := \log 3K$. Then

$$\|C(s)\| \leq M\,e^{\omega s}\ .$$

2.4. COROLLARY. For C as in 2.3. there are constants $M \geq 1$ and $\omega \geq 0$ with

$$\|C(t)\| \leq M \cdot \cosh\omega t \tag{4}$$

for all $t \in R$.

PROOF. Let $\|C(t)\| \leq M_1\,e^{\omega_1 t}$, $t \in R^+$, and put

$$M := 2M_1,\quad \omega := \omega_1\ .$$

Then

$$\| C(t) \| \leq \frac{M}{2} \cdot e^{\omega t} \leq \frac{M}{2} (e^{\omega t} + e^{-\omega t}) = M \cosh \omega t$$

for all $t \in R^+$.

2.5. REMARK. One is tempted to ask whether there exists a minimal ω satisfying (4) with some appropriate M . This is not true as can be seen from Example 2.2.2. Here with respect to the euclidean norm on R^2

$$\| C(t) \| = \sqrt{1 + \frac{1}{2} t^2 + \frac{1}{4} t^4}$$

and so for every $\omega > 0$ there is an $M_\omega \geq 1$ such that

$$\| C(t) \| \leq M_\omega \cdot \cosh \omega t$$

but C is not uniformly bounded in norm on R .

We are now going to introduce the concept of the infinitesimal generator A of a strongly continuous operator cosine function. Usually (see e.g. Sova [39]) $A:D(A) \to X$ is defined by

$$D(A) := \{ x \in X \mid t \to C(t)x \ \text{2-times differentiable in } 0 \}$$

$$Ax := C''(0)x \quad \text{for } x \in D(A) .$$

We shall instead begin by defining the resolvent operators of A instead of A itself.

Let C be a strongly continuous operator cosine function with $\| C(t) \| \leq M \cdot \cosh \omega t$, $t \in R$. Then for $z \in \mathbb{C}$ with $\mathrm{Re} z > \omega$

$$t \to e^{-zt} C(t)x$$

is Bochner integrable on $[0, \infty)$ for every $x \in X$. We put then

$$R(z)x := \frac{1}{z} \int_0^\infty e^{-zt} C(t)x \, dt , \quad x \in X_0 .$$

Then obviously

$$R(z) \in B(X)$$

and

$$R(z) C(t) = C(t) R(z) , \quad \mathrm{Re}\, z > \omega , \ t \in R .$$

Further we have

$$\|z\,R(z)\| \leq \int_0^\infty e^{-\mathrm{Re}\, z\cdot t}\|C(t)\|dt$$

$$\leq M \int_0^\infty e^{-\mathrm{Re}\, z\cdot t}\cosh \omega t \quad dt$$

$$= \frac{M}{2}\left(\frac{1}{\mathrm{Re}\; z-\omega} + \frac{1}{\mathrm{Re}\; z+\omega}\right)$$

and more generally for all $n \in \mathbb{N} \cup \{0\}$

$$\left\| \frac{d^n}{dz^n} z\,R(z) \right\| = \left\| \int_0^\infty t^n e^{-zt}\, C(t)\, dt \right\|$$

$$\leq \frac{Mn!}{2}\left[\frac{1}{(\mathrm{Re}\;\; z-\omega)^{n+1}} + \frac{1}{(\mathrm{Re}\;\; z+\omega)^{n+1}}\right].$$

2.6. THEOREM. There is a uniquely determined closed linear operator A on X such that for $z \in \mathbb{C}$ with $\mathrm{Re}\; z > \omega$

$$R(z) = R(z^2,A) := (z^2 I - A)^{-1}.$$

PROOF. We do not claim that A be densely defined! It is sufficient to show that $R(z)$ is injective and that it fulfils the resolvent equation (with respect to z^2!)

(5) $$(u^2-z^2)\, R(u)\, R(z) = R(z) - R(u).$$

The latter problem can be settled by a somewhat tedious but simple calculation.

Now let $R(z)x = 0$ for some $z \in \mathbb{C}$ with $\mathrm{Re}\; z > \omega$ and some $x \in X$. Then $R(u)\, R(z)x = 0$ for all $u \in \mathbb{C}$ with $\mathrm{Re}\; u > \omega$. According to (5) this leads to

$$R(u)x = 0$$

for all such u. Using the uniqueness theorem for the scalar Laplace transform applied to $(R(u)x|x')$, $x' \in X'$ (= continuous dual of X), we find that $C(t)x = 0$ for all $t \geq 0$, from which by 2.1. $x = 0$ follows.

Putting

$$A(z) := z^2 I - R(z)^{-1},$$

$A(z)$ is a closed operator on X with $D(A(z)) = R(R(z))$:= range of $R(z)$. This defines $A(z)$ independently of z if only $\mathrm{Re}\; z > \omega$: Let $y = R(z)x$, $x \in X$. Then

$$y = R(z)x = R(u)x + (u^2-z^2)\ R(u)\ R(z)x \in R(R(u))$$

for all $u \in \mathbb{C}$, $\operatorname{Re} u > \omega$.

So $D(A(z)) = R(R(z)) = R(R(u) = D(A(u))$ and for $x \in X$ we have

$$R(u)\ R(z)\ [z^2x - R(z)^{-1}x - u^2x + R(u)^{-1}x]$$

$$= (z^2-u^2)\ R(u)\ R(z)x - R(u)x + R(z)x = 0$$

and therefore the expression in brackets is equal to 0 which means

$$A(z)x = A(u)x\ .$$

So we may choose some z with $\operatorname{Re} z > \omega$ and put

$$A := A(z)\ .$$

2.7. DEFINITION. A as in 2.6. is called the infinitesimal generator of C.

2.8. THEOREM. With $R(z)$ defined as above

$$\lim_{\substack{z \in R \\ z \to \infty}} \| z^2\ R(z)x - x \| = 0$$

for each $x \in X$.

PROOF. Let $\varepsilon > 0$ be given and take $\delta > 0$ such that

$$\| C(t)x - x \| < \varepsilon$$

for all $0 < t < \delta$. For $z > \omega$ we have then

$$\| z^2\ R(z)x - x \| \leq z \int_0^\infty e^{-zs}\ \| C(s)x - x \|\, ds$$

$$\leq z\,\varepsilon \left(\int_0^\delta e^{-zs}\, ds\right) \| x \| + z\left(\int_\delta^\infty e^{-zs}(Me^{\omega s}+1)\, ds\right) \| x \|$$

$$\to \varepsilon \text{ for } z \to \infty.$$

This gives immediately

2.9. COROLLARY. The infinitesimal generator A of a strongly continuous operator cosine function C is a densely on X defined and closed linear operator.

In what follows A always denotes the infinitesimal generator of the strongly continuous operator cosine function C.

2.10. THEOREM. For all $t \in R$

$$C(t)\ D(A) \subset D(A)$$

and

$$C(t)\ Ax = AC(t)x$$

for $x \in D(A)$.

PROOF. If $x \in D(A)$ then there is a $y \in X$ such that for $\operatorname{Re} z > \omega$

$$x = R(z)y = \frac{1}{z} \int_0^\infty e^{-zs}\ C(s)y\ ds\ .$$

Now we have

$$C(t)x = C(t)\ R(z)y = R(z)\ C(t)y \in D(A)\ ,$$

and thus

$$AC(t)x = z^2\ C(t)x - C(t)\ R(z)^{-1}\ x$$

$$= z^2\ C(t)x - C(t)y\ .$$

On the other side also

$$C(t)\ Ax = z^2\ C(t)x - C(t)y\ .$$

2.11. THEOREM. For $x \in D(A)$

$$\lim_{t\to 0} \frac{1}{t^2} [C(-t)x-2C(0)x+C(t)x] = 2 \lim_{t\to 0} \frac{1}{t^2}\ [C(t)x-x]$$

exists and is equal to Ax.

PROOF. The difference quotients on both sides are equal since C is even.For $x \in D(A)$ we have $x = R(z)y$ for some $y \in X$ if $\operatorname{Re} z > \omega$. Then

$$C(t)x-x = C(t)\ R(z)y - R(z)y$$

$$= \frac{1}{z} \int_0^\infty e^{-zs}\ C(t)\ C(s)y\ ds - \frac{1}{z} \int_0^\infty e^{-zs} C(s)y\ ds\ .$$

From this equality the result follows by means of the functional equation and after some tedious calculations. For details see [20].

The symmetric difference quotients in 2.11. are not the most general second difference quotients to be used for the definition of the second derivative of C. Before treating this problem we prove

2.12. THEOREM. For $x \in D(A)$ and $t \in R$

(i) $\int_0^t (t-u)\, C(u)x\, du \in D(A)$

(ii) $$C(t)x-x = \int_0^t (t-u)\, C(u)\, Ax\, du = A \int_0^t (t-u)\, C(u)x\, du$$

PROOF. (i) Again for $x \in D(A)$ we can write

$$x = R(z)y$$

with $y \in X$ and $\operatorname{Re} z > \omega$. The function

$$u \to (t-u)\, C(u)y$$

is integrable on $[0,t]$, so

$$\int_0^t (t-u)\, C(u)y\, du \in X ,$$

$$\int_0^t (t-u)\, C(u)x\, du = \int_0^t (t-u)\, C(u)\, R(z)y\, du = \int_0^t (t-u)\, R(z)\, C(u)y\, du = R(z) \int_0^t (t-u)\, C(u)y\, du \in R(R(z)) = D(A).$$

(ii) Put $y = \int_0^t (t-u)\, C(u)x\, du$. Then according to 2.11. $\lim_{h\to 0} \frac{2}{h^2} (C(h)y-y) = Ay$. On the other hand as is shown in Sova [39], p. 14-16,

$$\lim_{h\to 0} \frac{2}{h^2} (C(h)y-y) = C(t)x - x .$$

This provides us with

$$C(t)x-x = A\int_0^t (t-u)\, C(u)x\, du .$$

Approximating A again by divided differences and using the fact that all the $C(t)$ commute gives then

$$C(t)x-x = \int_0^t (t-u)\, C(u)\, Ax\, du .$$

2.13. REMARK. 2.12. was generalized to Taylor's formula by Nagy [34].

2.14. THEOREM. Let $x \in D(A)$. Then

(i) $t \to C(t)x$ is two times differentiable on R.

(ii) $\frac{d}{dt} C(t)x \to 0$ for $t \to 0$.

(iii) $\frac{d^2}{dt^2} C(t)x = C(t)Ax = AC(t)x$ for every $t \in R$.

PROOF. By 2.12. we have

$$C(t)x = x + \int_0^t (t-u)\, C(u)\, Ax\, du$$

$$= x + t\int_0^t C(u)\, Ax\, du - \int_0^t uC(u)\, Ax\, du ,$$

so

$$C'(t)x = \int_0^t C(u)\, Ax\, du ,$$

$$\| C'(t)x \| \leq |t|\, M\, e^{\omega|t|} \|Ax\| \to 0 \quad \text{for } t \to 0 .$$

Differentiating once more we get

$$C''(t)x = C(t)\, Ax .$$

2.15. REMARK. There are three types of "infinitesimal generators" involved:

$$D(A) = R(R(z)) , \quad \text{Re } z > \omega ,$$

$$A = z^2 I - R(z)^{-1} ;$$

$$D(B) = \{x \in X \mid \lim_{h\to 0} \frac{2}{h^2}(C(h)x-x) \text{ exists}\}$$

$$Bx = \lim \frac{2}{h^2}(C(h)x-x) ;$$

$$D(C) = \{x \in X \mid t \to C(t)x \text{ 2-times diff. on } R\}$$

$$Cx = C''(0)x .$$

We have shown above that $A \subset B$ and $A \subset C$. In fact it is true that $A = B = C$ (see Sova [39], Travis & Webb [42], Lutz [20]).

2.16. EXAMPLES. We refer to the examples in 2.2. Therein:

1. $A = a$

2. $A = \begin{pmatrix} 0 & 0 \\ 1 & 0 \end{pmatrix}$

3. $A = \frac{d^2}{ds^2}$

According to 2.6. the resolvent operators of A are given by means of the Laplace transform of the associated operator cosine function. We can also apply the reverse Laplace transform ([25]) :

2.17. THEOREM. If C satisfies $\|C(t)\| \leq Me^{\omega|t|}$, $t \in R$, for $a > \omega$ and every $x \in X$

$$C(t)x = \frac{1}{2\pi i} \int_{a-i\infty}^{a+i\infty} e^{zt}\, z\, R(z^2,A)x\, dz \quad .$$

We end this chapter by adding some remarks on uniformly continuous operator cosine functions. If $A \in B(X)$

$$C(t) := \cosh t\sqrt{A} := \sum_{n=0}^{\infty} \frac{t^{2n}}{(2n)!} A^n$$

defines a uniformly continuous operator cosine functions ; A is the infinitesimal generator of C. That this is already the most general case is the result stated in

2.18. THEOREM. If $C: R \to B(X)$ is a uniformly continuous operator cosine function then there is an $A \in B(X)$ with

$$C(t) = \cosh t\sqrt{A} \quad , \qquad t \in R .$$

We have $A = \lim_{s \to 0} \frac{2}{s^2}(C(s)-I)$ in the uniform operator topology.

PROOF [23] : For $f: \mathbb{C} \to \mathbb{C}$, $f(z) := \cosh\sqrt{z}$, we have $f'(0) = \frac{1}{2} \neq 0$. Thus there is locally near $f(0) = 1$ a holomorphic inverse of f, say g. g is then holomorphic in an open neighborhood of $\sigma(C(0)) = \{1\}$ and hence also of $\sigma(C(t))$ for all $|t| < \varepsilon$ with some $\varepsilon > 0$.

For $|t| < \varepsilon$

$$H(t) := g(C(t)) \in B(X)$$

is defined in the sense of the holomorphic functional calculus. Take t_o with $|t_o| < \varepsilon$ and put

$$A: = \frac{1}{t_o^2} H(t_o) .$$

Then $f(t_o^2 A) = C(t_o)$.

Now we use the fact that each (scalar or operator valued) solution K of d'Alembert's functional equation 2.1. (i), (ii) satisfies for $n \in N$

$$(6) \qquad K(nt) = a_o I + a_1 K(t) + \ldots + a_n K(t)^n, \qquad t \in R ,$$

where the integers $a_o, \ldots, a_n$ depend upon n but are independent of K . So for each $n \in N$.

$$\begin{aligned} f(n^2 t_o^2 A) &= a_o I + \ldots + a_n \left[f(t_o^2 A)\right]^n \\ &= a_o I + \ldots + a_n C(t_o)^n \\ &= C(nt_o) \end{aligned}$$

and thus

$$u^2 t_o^2 A = H(n t_o)$$

if only $|n t_o| < \varepsilon$.

Analogously for each rational $\frac{m}{n}$ with $|\frac{m}{n}| < 1$ we see that $\frac{m^2}{n^2} t_o^2 A = H(\frac{m}{n} t_o)$. But then the uniform continuity of C gives

$$t^2 t_o^2 A = H(t t_o) \quad \text{for } |t| < 1 ,$$

and thus

$$t^2 A = H(t) \qquad \text{for } |t| < 1 ,$$

which gives

$$C(t) = f(H(t)) = \cosh t\sqrt{A} \quad \text{for these } t .$$

By means of (6) this equality extends to all of R .
We have

$$\lim 2 \cdot \frac{(\cosh s \sqrt{z}) - 1}{s^2} = 2 \lim_{s \to 0} \sum_{n=1}^{\infty} \frac{s^{2n-2} z^n}{(2n)!} = z$$

uniformly with respect to z on compact subsets of $\mathbb{C}$. Hence the holomorphic functional calculus $(z \to A)$ gives also the second assertion.

2.19. REMARK. Another proof along the lines of the analogous reasoning in the scalar valued case was given by Kurepa [19].

3. THE LINEAR SECOND ORDER CAUCHY PROBLEM

In this chapter we present only the most basic facts concerning the connections between operator cosine functions and linear second order differential equation with values in Banach spaces. More detailed information, especially dealing with the problem of reduction to a first order system, can be found e.g. in Fattorini [5], Travis & Webb [43],[44], Kisynski[12],[13],[14].

3.1. DEFINITION. Let A denote a closed and densely defined linear operator on X. We say that the Cauchy problem for the linear differential equation

$$\frac{d^2}{dt^2} u = Au$$

is uniformly well-posed iff

(i) For each pair $x,y \in D(A)$ there is exactly one 2-times continuously differentiable function

$$u:R \to D(A)$$

such that

$$\begin{aligned} u''(t) &= Au(t) \ , \quad t \in R \\ u(0) &= x \ , \\ u'(0) &= y \ . \end{aligned} \tag{7}$$

(ii) u as in (i) depends continuously upon (x,y) with respect to uniform convergence on compact subsets of R.

3.2. DEFINITION. If C is a strongly continuous cosine function on X, then $S:R \to B(X)$ defined by

$$S(t)x := \int_0^t C(s)x\, ds \ , \qquad x \in X \ ,$$

is called the associated sine function. Obviously $S(0) = 0$.

3.3. THEOREM. Let A be the infinitesimal generator of the strongly continuous operator cosine function C and let S denote the sine function associated with C. Then:

(i) The Cauchy problem for $\frac{d^2}{dt^2}u = Au$ is uniformly well-posed.

(ii) For $x,y \in D(A)$ and a continuously differentiable $g:R \to X$ the only solution of

$$u''(t) = Au(t) + g(t) \quad ,$$
$$(8) \qquad u(0) = x \; ,$$
$$u'(0) = y \; .$$

is given by

$$(9) \qquad u(t) = C(t)x + S(t)y + \int_0^t S(t-s)\, g(s)\, ds \; , \quad t \in R \; .$$

PROOF. α) u as in (9) fulfils (8): For $x,y \in D(A)$ we have

$$(10) \qquad C''(t)x = AC(t)x \; ;$$

and it is easy to see that

$$(11) \qquad S''(t)y = C'(t)y = AS(t)y \; .$$

Moreover

$$(12) \qquad \frac{d^2}{dt^2}\left(\int_0^t S(t-s)g(s)ds\right) = \frac{d}{dt}\left(\int_0^t C(t-s)g(s)ds + S(0)g(t)\right)$$
$$= \frac{d}{dt}\left(\int_0^t C(t-s)g(s)ds\right)$$
$$= A\left(\int_0^t S(t-s)g(s)ds\right) + C(0)\, g(t)$$
$$= Au(t) + g(t) \quad ,$$

where one has got to make sure that $\int_0^t S(t-s)g(s)ds \in D(A)$. Since u satisfies obviously the given initial values, (10),(11),(12) together show that u fulfils (8).

β) Let u be a solution of (8) with $u:R \to D(A)$. Then

$$\frac{d}{ds} S(t-s)\, u'(s) = -C(t-s)\, u'(s) + S(t-s)\, u''(s) \; ,$$

$$\frac{d}{ds} S(t-s)\, u(s) = -S(t-s)\, Au(s) + C(t-s)\, u'(s) \; .$$

Integrating on both sides over $[0,t]$ and taking into account that $C(0) = I$ and $S(0) = 0$ gives

$$-S(t)y = -S(t)\,u'(0)$$

$$= -\int_0^t C(t-s)\,u'(s)ds + \int_0^t S(t-s)\,u''(s)\,ds\ ,$$

$$u(t) - C(t)x = u(t) - C(t)\,u(0)$$

$$= -\int_0^t S(t-s)\,Au(s) + \int_0^t C(t-s)\,u'(s)\,ds\ .$$

Inserting $u''(s) = Au(s) + g(s)$ and adding up on both sides gives (9).

γ) If two sets (x_o, y_o), (x_1, y_1) of initial values are taken from $D(A)$ then the respective solutions of (8) (with g=0) are given by

$$u_o(t) = C(t)x_o + S(t)y_o\ .$$

and

$$u_1(t) = C(t)x_1 + S(t)y_1\ .$$

Thus,

$$\|u_o(t) - u_1(t)\| \leq \|C(t)\|\cdot\|x_o - x_1\| + \|S(t)\|\cdot\|y_o - y_1\|\ .$$

Since there are bounds $M \geq 0$ and $\omega \geq 0$ with

$$\|C(t)\| \leq Me^{\omega|t|}$$

and a fortiori

$$\|S(t)\| \leq M\,|t|\,e^{\omega|t|}$$

the continuous dependence of u upon the initial values in the above mentioned sense follows.

3.4. REMARKS. 1. Part β) in the proof of 2.3. shows that any solution of (7) for continuous g is of the form (8).

2. Fattorini [5] has shown that the uniform well-posedness of the Cauchy problem for $\frac{d^2}{dt^2}u = Au$ is in fact equivalent to A being the infinitesimal generator of a strongly continuous operator cosine function.

It is quite natural to ask whether the second order initial value problem (7) can always be reduced to a uniformly well-posed first order problem on some subspace of $X \times X$ containing $D(A) \times X$. This is indeed always possible using the familiar reduction

$$\begin{pmatrix} 0 & I \\ A & 0 \end{pmatrix}$$

with respect to another norm the definition of which uses esplicit bounds for the cosine function generated by A (see Kisynski [14]). With respect to the original norm uniform well-posedness can get lost in the reduction procedure.

If A is densely defined and has a square root $A^{1/2}$ (in the usual sense) which generates a strongly continuous operator group G on R then A is the infinitesimal generator of the cosine function C given by

$$C(t) = \frac{1}{2}(G(t) + G(-t)) \tag{13}$$

For this to be true Fattorini [5] gave sufficient conditions. But by examples given by Kisynski [12],[13] a representation of the form (13) of a strongly continuous operator cosine function C does not exist in general.

So it seems to be appropriate to develop separate though somewhat similar theories for first and second order problems by means of semigroups resp. cosine functions. Of course one could ask whether one has got to establish separate theories for all orders. Fortunately, this is not true. We have the following theorem due to Fattorini [5] :

3.5. THEOREM. Let A be a closed and densely defined linear operator on X . Then the Cauchy problem for the X-valued differential equation

$$\frac{d^n}{dt^n} u = Au \qquad (n > 2) \quad \text{is uniformly well-posed}$$

(a term which is defined mutatis mutandis as in 3.1.) iff

(i) A is bounded

and

(ii) $\sum_{j=0}^{\infty} \frac{t^j}{(u_j)!} A^j$ converges in the norm topology of $B(X)$.

4. CHARACTERIZATION OF INFINITESIMAL GENERATORS AND PERTURBATION THEORY

Having established the connection between operator cosine functions and second order Cauchy problems it is quite natural to ask for an intrinsic characterization of infinitesimal generators of strongly continuous operator cosine functions.

Such a criterian was given by Sova[39] and Fattorini[5]. It is closely related to the celebrated Hille-Yoshida theorem on generators of operator semi-groups.

4.1. THEOREM. Let A be a closed and densely defined linear operator on X. Then A is the infinitesimal generator of a strongly continuous operator cosine function with

$$\| C(t) \| \leq M \cos \omega t \ , \quad t \in \mathbb{R} \ ,$$

iff for all $z \in \mathbb{C}$ with $\operatorname{Re} z > \omega$

(i) $\quad z^2 \in \rho(A)$

(ii) $\quad \left\| \frac{d^n}{dz^n} zR(z^2,A) \right\| \leq \frac{M \cdot n!}{2} \left\{ \frac{1}{(\operatorname{Re} z-\omega)^{n+1}} + \frac{1}{(\operatorname{Re} z+\omega)^{n+1}} \right\}$

The proof for the necessity of these conditions was given in our second chapter above.

The spectrum of the infinitesimal generator of C is thus contained in a parabolic domain in $\mathbb{C}$ open to the left. If A is the infinitesimal generator of a strongly continuous operator cosine function it generates also a strongly continuous semi-group of operators. The converse of this statement is obviously not true since there are semi-group generators whose spectrum contains a left half plane in $\mathbb{C}$.

There is something seriously unpleasant about Sova's criterion 4.1. The analogous criterion for semi-group uses bouds of the form

$$n! \, \| R(z,A)^{n+1} \| = \left\| \frac{d^n}{dz^n} R(z,A) \right\| \leq Mn! \, \frac{1}{(\operatorname{Re} z-\omega)^{n+1}} \ , \ \operatorname{Re} z > \omega .$$

Usually one studies contraction semi-groups what means the case $M = 1$, $\omega = 0$. So all one has got to show for proving that A is the generator of a contraction semi-group is

$$\sigma(A) \subset \{z \mid \operatorname{Re} z \leq 0\} ,$$

$$\|R(z,A)\| \leq \frac{1}{\operatorname{Re} z} , \quad \operatorname{Re} z > 0 ,$$

The situation is much more involved even in the case $M = 1, \omega = 0$ because one has to establish bounds for $\frac{d^n}{dz^n} zR(z^2,A)$. An expansion of these expressions in terms of $R(z^2,A)$ is given in

4.2. THEOREM [30]. If $n = 2m$, $m \in N_o$, then

$$\frac{d^n}{dz^n} zR(z^2,A) = \sum_{i=m}^{n} (-1)^i \frac{(n+1)!\,i!}{(n-i)!\,(2i-n+1)!} z \cdot (2z)^{2i-n} R(z^2,A)^{i+1} .$$

If $n = 2m+1$, $m \in N_o$, then

$$\frac{d^n}{dz^n} zR(z^2,A) = \sum_{i=m+1}^{n} (-1)^i \frac{(n+1)!\, i!}{(n-i)!\,(2i-n+1)!} z(2z)^{2i-n} R(z^2,A)^{i+1}$$
$$+ (-1)^m n!\, R(z^2,A)^{m+1} .$$

It is easily seen that it is not sufficient to assume bounds on $R(z^2,A)$ alone.

To put it in practical terms: Sova's criterion cannot be used directly to prove that a given operator is an infinitesimal generator. So it seems to be quite natural to ask for perturbation methods circumventing Criterion 4.1.

We start with a class of operators for which it is easily seen by means of functional calculus that they are generators.

4.3. THEOREM. A normal operator A with spectral measure E on a Hilbert space X is the infinitesimal generator of a strongly continuous operator cosine function C iff there is an $\omega \in R$ with

$$\operatorname{Re} z > \omega \Rightarrow z^2 \in \rho(A) .$$

In this case C is given by

$$C(t)x = \int_{\mathbb{C}} \cosh \sqrt{\lambda}\, t \; d\,E(\lambda)x , \quad x \in X .$$

PROOF. Under our assumptions the function $\lambda \to \cosh \sqrt{\lambda}\, t$ is bounded on $\sigma(A)$ for each $t \in R$.

Nagy [34] proved in 1974 that under bounded perturbations the property of being a generator is presented. This result is generalized in the next theorem.

We discuss the uniform well-posedness of the Cauchy problem for

$$\frac{d^2u}{dt^2} = A(t)u \ , \quad t \in R \ ,$$

where the time-dependent operator $A(t)$ can be split according to

$$A(t) = A + B(t), \quad t \in R \ ,$$

A being the infinitesimal generator of a cosine function and $B(t) \in B(X)$ for all $t \in R$.

We state the theorem in a form that indicates the main steps of the proof ([27]).

4.4. THEOREM. Let A be the infinitesimal generator of the operator cosine function C_o with

$$\|C_o(t)\| \leq M e^{\omega|t|} \ , \ t \in R \ ,$$

and let S_o be the sine function associated with C_o. Let $B: R \to B(X)$ be 2-times strongly continuously differentiable. Then:

1. The Cauchy problem for the X-valued differential equation

$$(14) \qquad \frac{d^2}{dt^2} u(t) = (A + B(t)) u(t), \quad t \in R \ ,$$

is uniformly well-posed.

2. The solution of (14) with

$$u(0) = x$$
$$u'(0) = y$$

for $x,y \in D(A)$ is given by

$$u(t) = C(t)x + S(t)y \ ,$$

where $C,S: R \to B(X)$ are strongly continuous, $t \to C(t)x$ and $t \to S(t)x$ are for $x \in D(A)$ 2-times continuously differentiable solutions of (14) fulfilling

$$C(0)x = x \ , \qquad C'(0)x = 0 \ ,$$
$$S(0)x = 0 \ , \qquad S'(0)x = x \ .$$

3. C and S are given by

$$C(t)x = C_o(t)x + \sum_{n=1}^{\infty} C_n(t)x \quad ,$$

$$S(t)x = S_o(t)x + \sum_{n=1}^{\infty} S_n(t)x \ , \quad x \in X \ ,$$

where

$$C_n(t)x := \int_0^t S_o(t-s)\ B(s)\ C_{n-1}(s)x\ ds \quad ,$$

(15)

$$S_n(t)x := \int_0^t S_o(t-s)\ B(s)\ S_{n-1}(s)x\ ds, \ n \in N, \quad x \in X \ .$$

4. With $K_s := \sup_{\substack{t \in [0,s] \\ (t \in [s,0])}} \{\|B(t)\|, \|B'(t)\|\}$, $s \in R$,

we have for all $t \in [0,s]$ (or $t \in [s,0]$) the bounds

$$\|C(t)\| \leq M\ e^{\omega|t|} \quad (\cosh \sqrt{MK_s}\ t) \quad ,$$

$$\|S(t)\| \leq M\ e^{\omega|t|} \quad (\frac{1}{\sqrt{MK_s}} \sinh \sqrt{MK_s}\ t\ ) \ ,$$

$$\|C(t) - C_o(t)\| \leq M\ e^{\omega|t|} (\cosh \sqrt{MK_s} t) -1) \ ,$$

$$\|S(t) - S_o(t)\| \leq M\ e^{\omega|t|} ((\frac{1}{\sqrt{MK_s}} \sinh \sqrt{MK_s}\ t)-1) \ ,$$

where for $r,t \in R$

$$\cosh \sqrt{r}t := \sum_{n=0}^{\infty} \frac{r^n t^{2n}}{(2n)!} \quad ,$$

$$\frac{1}{\sqrt{r}} \sinh \sqrt{r}t := \sum_{n=0}^{\infty} \frac{r^n t^{2n+1}}{(2n+1)!} \quad .$$

4.5. REMARK. Travis & Webb have in [45] proved ananalogous result for time-independent perturbations B such that

(i) B is closed and densely defined,

(ii) $S_o(t)\ X \subset D(B)$

(iii) $t \to BS_o(t)x$ is continuous for all $x \in X$.

They use instead of (15) the recursion formula

$$C_n(t)x := \int_0^t C_o(t-s)\ B\ S_{n-1}(s)x\ ds$$

which is found integrating (formally) by parts in (15).

There are more results on perturbation of operator cosine functions by Travis & Webb, Takenaka & Okazawa, Fattorini, Lutz et.al.

5. SPECTRAL THEORY

The information in this chapter are mainly taken from Nagy [34], Lutz [23], [24], Giusti [8] .

Let C denote a strongly continuous operator cosine functions on X with infinitesimal generator A .

As easy consequences of d'Alembert's functional equation, and the spectral mapping theorem for bounded linear operators we have

5.1. THEOREM.

(i) For no $t \in R$ $C(t)$ and $C(2t)$ are quasinilpotent.

(ii) If $\dim X = \infty$, for no $t \in R$ $C(t)$ and $C(2t)$ are compact.

(iii) If $\sigma(C(t)) \subset R$ (resp. R^+) for all $t \in R$ then $\sigma(C(t)) \subset [-1,\infty)$ (resp. $[1,\infty)$) for all $t \in R$.

Here, $R^+ := [0,\infty)$, $R^- := (-\infty,0]$.

To relate information concerning $\sigma(A)$ to $\sigma(C(t))$ and vice versa we need a spectral mapping theorem. We state first following Nagy [34]:

5.2. LEMMA. For $s \in R$ and $z \in \mathbb{C}$

$$S(s,z)x := \int_0^s \sinh z(s-t)\, C(t)x\, dt , \quad x \in X ,$$

defines a bounded linear operator $S(s,z)$ on X such that for all $x \in X$

$$(A-z^2 I)\, S(s,z)x = z(C(s)-(\cosh z\, s)I)\, x . \tag{16}$$

For $x \in D(A)$

$$(A-z^2 I)\, S(s,z)x = S(s,z)(A-z^2 I)x . \tag{17}$$

5.3. THEOREM. For $s \in R$ $\cosh s\sqrt{\sigma(A)} \subset \sigma(C(s))$.

PROOF. We use the following decomposition of $\sigma(B)$ for a closed and densely defined operator B

$$\sigma_1(B) := \{\lambda \in \mathbb{C} \mid \lambda I - B \quad \text{not injective}\}$$

$$\sigma_2(B) := \{\lambda \in \mathbb{C} \mid R(\lambda I - B) \text{ not dense in } X\}$$

$$\sigma_3(B) := \{\lambda \in \mathbb{C} \mid \exists \{x_n\} \subset D(B), \|x_n\| = 1, \lim \|(\lambda I - B)x_n\| = 0\}$$

Then $\sigma(B) = \sigma_1(B) \cup \sigma_2(B) \cup \sigma_3(B)$ where contrary to the usual definition of the point, residual, and continuous spectrum the sets σ_i need not be disjoint.

For $s = 0$ our assertion is trivially true. So let $s \neq 0$. Then for $z \neq 0$, by means of (16) and (17), $z^2 \in \sigma_1(A)$ implies that $C(s)x = \cosh(zs)x$ for some $0 \neq x \in D(A)$; that means $\cosh zs \in \sigma_1(C(t))$. Likewise for $z \in \sigma_i(A)$, $i = 2,3$, $\cosh \sqrt{z}s \in \sigma_i(C(s))$.

To $z = 0$ an analogous reasoning applies when we use instead of (16) and (17) the equality

$$A \int_0^s (s-t)C(t)x\,dt = \int_0^s (s-t)C(t)\,Ax\,dt = C(t)x - x, \quad x \in D(A),$$

established in 2.12.

5.4. COROLLARY. If $\sigma(C(t)) \subset \mathbb{R}$ $(\mathbb{R}^+)$ for all $t \in \mathbb{R}$, then also

$$\sigma(A) \subset \mathbb{R} \ (\mathbb{R}^+).$$

If C is periodic, that is if $C(t+s) = C(t)$, $t \in \mathbb{R}$, for some $s > 0$, C is uniformly bounded in norm and therefore $\sigma(A) \subset \mathbb{R}^-$. A more detailed spectral characterization of periodic cosine functions is given in

5.5. THEOREM. C is 2π-periodic iff the following three conditions are satisfied

(i) $\sigma(A) \subset -\mathbb{N}^2 \cup \{0\}$

(ii) $\sigma(A)$ contains only simple poles of $R(z,A)$

(iii) $\sum_{m \in \mathbb{N}_0} P_m x = x$ for all $x \in D(A)$,
where P_m denotes the projection associated with the spectral set $\{-m^2 \mid m \in \mathbb{N}_0\}$.

5.6. COROLLARY. C is 2π-periodic $\Longleftrightarrow C(2\pi) = I$. If C is periodic $\sigma(A)$ is bounded iff A is bounded.

Likewise, for uniformly bounded C on a Hilbert space X , A is bounded iff $\sigma(A)$ is bounded. This follows from the fact that by a result of Fattorini [6] such a cosine function can be jointly transformed to a strongly continuous cosine function of selfadjoint operators. Recently the following result was found ([29]).

5.7. THEOREM. Let C be uniformly bounded. Then

(i) $\sigma(A) \neq \emptyset$, if $X \neq \{0\}$;

(ii) $\sigma(A)$ is bounded iff A is bounded (that is the same as stating that C is uniformly continuous).

<u>Remarks to the bibliography</u>. We did not strive for completeness and have listed only papers dealing directly with our subject. So we did not include papers dealing mainly with algebraic questions , scalar functional equations, related functional equations, and refrained also from listing texts handling second-order differential equations by means of a reduction to a first order system which is usually done by assuming the existence of square roots of certain operators.

REFERENCES

[1] Baker,J.A,,D'Alembert's functional equation in Banach algebras, Acta Sci.Math. Szeged <u>32</u>, (1971),225-234.

[2] Baker,J.A. Davidson,K.R.: Cosine exponential and quadratic functions, Preprint.

[3] Buche,A.B., On the cosine - sine operator functional equations, Aequ. Math. <u>6</u>, (1971), 231-234.

[4] Da Prato,G.,Giusti,E., Una catterizzatione dei generatori di funzioni coseno astratto, Boll.Unione Mat.Ital.<u>22</u> (1967), 357-362.

[5] Fattorini,H.O., Ordinary differential equations in linear topological spaces, I,II, J.Diff.Equ. <u>5</u> (1968),72-105, <u>6</u>(1969) 50-70.

[6] Fattorini,H.O., Uniformly bounded cosine functions in Hilbert spaces, Indiana Univ. Math. J. <u>20</u> (1070), 411-425.

[7] Faulkner,G.D., Shonkwiler,R.W., Cosine representations of abelian *-semigroups and generalized cosine operator functions, Can.J.Math. <u>30</u> (1978), 474-482.

[8] Giusti,E., Funzioni coseno periodiche, Boll.Unione Mat.Ital.<u>22</u> (1967), 478-485.

[9] Goldstein,J.A., On the convergence and approximation of cosine functions, Aequ.Math. <u>10</u> (1974), 201-205.

[10] Goldstein,J.A., On a connection between first and second order differential equations in Banach spaces, J.Funct.Anal. <u>4</u>(1969),50-70.

[11] Goldstein,J.A., Radin,Ch., Showalter,R.E., Convergence rates of ergodic limits for semigroups and cosine functions,

Semigroup Forum 16(1978), 89-95.

[12] Kisynski,J., On operator-valued solutions of d'Alembert's functional equation I , Coll. Math. 23 (1971),107-114.

[13] Kisynski,J., On operator-valued solutions of d'Alembert's functional equations II, Studia Math. 24 (1972), 43-66.

[14] Kisynski, J., On cosine operator functions and one-parameter groups of operators, Studia Math.44 (1972),93-105.

[15] Konishi,Y., Cosine functions of operators in locally convex spaces, J.Fac.Sci.Univ.Tokyo Math. 18, (1971/72),443-463.

[16] Kurepa, S., A cosine functional equation in n-dimensional vector space, Glasnik Mat. 13 (1958), 169-189.

[17] Kurepa, S., A cosine functional equation in Hilbert space, Can. J.Math. 12 (1960) , 45-50.

[18] Kurepa, S., On some functional equations in Banach space, Studia Math. 19 (1960), 149-158.

[19] Kurepa, S., A cosine functional equtaion in Banach algebras, Acta Sci.Math. Szeged 23 (1962), 255-267.

[20] Lutz, D., Der infinitesimale Erzeuger für Lösungen der Funktionalgleichung des Cosinus mit Werten im Bereich der abgeschlossenen linearen Operatoren auf einem Banachraum, Seminarberichte FB Mathematik 4 (1978), 43-70. FU Hagen.

[21] Lutz, D., Which operators generate cosine operator functions? Rendic.Accad.Naz.Lincei 63(1978), 314-317.

[22] Lutz, D., Compactness properties of operator cosine functions, C.R.Math.Rep.Acad.Sci.Canada 2 (1980), 277-280.

[23] Lutz, D., Über operatorwertige Lösungen der Funktionalgeichung des Cosinus, Math.Z. 171 (1980) 233-245.

[24] Lutz, D., Periodische operatorwertige Cosinusfunktionenen,Resultate der Mathematik 4, (1981), 75-83.

[25] Lutz, D., Representation of operator-valued cosine functions by Laplace transform, Preprint.

[26] Lutz, D., Über die Konvergenz operatorwertiger Cosinusfunktionen mit gestörtem infinitesimalen Erzeuger, Periodic.Math. Hung. (to appear).

[27] Lutz, D., On bounded time-dependent perturbations of operator cosine functions, Aequ. Math. (to appear).

[28] Lutz, D., Zur Approximation operatorwertiger Cosinusfunktionen, Preprint.

[29] Lutz, D., Some spectral properties of bounded operator cosine functions, Preprint.

[30] Lutz, D., On resolvents of generators of operator cosine functions, Preprint.

[31] Maltese,G., Spectral representations for solutions of certain abstract functional equations, Comp.Math. 15 (1962), 1-22.

[32] Maltese,G., Spectral representations for some unbounded normal operators, Trans. Amer.Math.Soc.110(1964), 72-87.

[33] Nagy,B., On the generators of cosine operator functions, Publ. Math. 21 (1974), 151-154.

[34] Nagy, B., On cosine operator functions on Banach spaces, Acta Sci. Math. Szeged 36 (1974), 281-290.

[35] Nagy, B., Cosine operator functions and the abstract Cauchy problem, Period.Math. Hung. 7 (1976), 213-217.

[36] Nagy, B., Approximation theorems for cosine operator functions, Acta Math.Acad.Sci.Hung. 29 (1977), 69-76.

[37] Nelson,S., Triggiani,R., Analytic properties of cosine operators, Proc.Amer.Math.Soc. 74 (1979), 101-104.

[38] Rankin III,S.M., A remark on cosine families, Proc.Amer.Math.Soc. 79(1980), 376-378.

[39] Sova, M., Cosine operator functions, Rozprawy Mathematyczne 49 Warszawa 1966.

[40] Sova, M., Semigroups and cosine functions of normal operators in Hilbert space, Casopis pest.mat. 93,(1968),437-458.

[41] Takenaka,T., Okazawa,N., A Phillips-Miyadera type perturbation theorem for cosine functions of operators, Tôhoku Math. J. 30 (1978), 107-115.

[42] Travis,C.C., Webb,F.F., Compactness, regularity and uniform continuity properties of strongly continuous families, Houston J.Math. 3 ,(1977), 555-567.

[43] Travis,C.C., Webb.G.F., Cosine families and abstract nonlinear second order differential equations, Acta Math. Acad. Sci.Hung. 32 (1978), 75-96.

[44] Travis, C.C., Webb,G.F., An abstract second order semilinear Volterra integrodifferential equation, SIAM J.Math.Anal.10 (1979), 412-424.

[45] Travis,C.C., Webb.G.F., Perturbation of strongly continuous cosine family generators, Preprint.

[46] Vajzović,F., Einige Funktionalgleichungen im Fréchetschen Raum, Glasnik Mat. 3 (1968), 19-38.

RANK AND INDEX IN BANACH ALGEBRAS

Hrvoje Kraljević

INTRODUCTION

The aim of this lecture is to consider certain natural generalizations of some well-known notions from the theory of operators on Banach spaces to the case of general Banach algebras. It will be explained how to generalize the notions of finite rank operators, compact operators, Fredholm operators and the index of a Fredholm operator.

In Section 1 we shall recall the definitions and consider some properties of these classical notions, and explain where the ideas for generalizations came from.

In Section 2 we define the notion of rank in general Banach algebras and derive some of its properties.

Section 3 is devoted to the definitions of compact and Fredholm elements in Banach algebras and of an index function on the set of Fredholm elements. The properties of this index are stated only as a conjecture.

In Section 4 we consider in detail the properties of finite rank elements in a semisimple Banach algebra and prove the conjecture of Section 3 in this case.

In Section 5 we show in an example that our index function does not bear all the properties of the classical index. To eliminate this failure we give a new (in a sense "finer") definition of index in the case of semiprime Banach algebras and conjecture its properties. The conjecture was proved in [9] in the special case of semisimple Banach algebras.

Most of the results were obtained in [8],[9] and [12]. However, the approach in the present article is more direct and the framework of the theory is more general.

Let us fix some notation. If X,Y are Banach spaces, $L(X,Y)$ denotes the Banach space of bounded linear operators $X \to Y$, and $L(X) = L(X,X)$. For an algebra A with the identity $G(A)$ denotes the group of invertible elements in A, $P(A)$ is the set of all projections in A (i.e. elements $p \in A$ with $p^2 = p$) and $\sigma(a)$ is the spectrum of $a \in A$.

1. SOME NOTIONS IN CLASSICAL OPERATOR THEORY

Let X be a complex Banach space. An operator $A \in L(X)$ is called of finite rank if $\dim AX$ is finite ; then $r(A) = \dim AX$ is called the rank of A . Let $D(X)$ be the set of all finite rank operators in $L(X)$. Then $D(X)$ is a two-sided ideal in $L(X)$; note that $D(X)$ is not closed unless X is finite-dimensional.

For $A \in L(X)$ denote by $L(A)$ the subalgebra of $L(X)$ generated by A :

$$L(A) = \text{span} \{ A, A^2, A^3, \ldots \} .$$

LEMMA 1.1. For any $A \in D(X)$ the algebra $L(A)$ is finite-dimensional. More precisely, $\dim L(A) \leq r(A)$.

PROOF. Let $Z = AX$ and $A_1 = A|Z \in L(Z)$. Z being of finite dimension $n = r(A)$, there exist $a_0, \ldots, a_{n-1} \in C$ such that

$$A_1^n = a_0 I + a_1 A_1 + \ldots + a_{n-1} A_1^{n-1} ,$$

where I is the identity operator on Z . Now,

$$A^{n+1} = A_1^n A = a_0 IA + a_1 A_1 A + \ldots + a_{n-1} A_1^{n-1} A =$$
$$= a_0 A + a_1 A^2 + \ldots + a_{n-1} A^n .$$

This shows that $L(A)$ is spanned by $A, A^2, \ldots, A^n$, therefore $\dim L(A) \leq n$. □

Since obviously $r(BAC) \leq r(A)$, we have that $\dim L(BAC) \leq r(A)$ for any $B, C \in L(X)$. The following theorem shows that the rank of A is exactly the supremum of these dimensions. This will allow the "algebraization" of the notion of rank.

THEOREM 1.2. Let $A \in D(X)$. Then $r(A) = \sup \{ \dim L(AB) : B \in L(X) \} = \sup \{ \dim L(BA) : B \in L(X) \} = \sup \{ \dim L(BAC) : B, C \in L(X) \}$. If $A \in L(X) \setminus D(X)$ then the three supremums are infinite.

PROOF. Since for any $B,C \in L(X)$ $\dim \mathcal{L}(BAC) \leq r(A)$ and since obviously the first and the second supremum are bounded above by the third one, to prove the first assertion we have to show that there exist $B,C \in L(X)$ such that $\dim \mathcal{L}(BA) = \dim \mathcal{L}(AC) = r(A)$.

Let Y be a subspace of X such that $X = Y \dotplus \ker A$. Then Y is $r(A)$-dimensional and $A_1 = A|Y \in L(Y,AX)$ is an isomorphism. Let $B_1, C_1 \in L(AX,Y)$ be isomorphisms such that $B_1A_1 \in L(Y)$ and $A_1C_1 \in L(AX)$ generate $r(A)$-dimensional algebras (e.g. such that B_1A_1 and A_1C_1 both have simple non-zero spectra). Let Z be a closed subspace of X such that $X = Z \dotplus AX$. Define $B,C \in L(X)$ by $B|AX = B_1$, $B|Z = 0$, $C|AX = C_1$, $C|Z = 0$. Then, obviously, $\dim \mathcal{L}(BA) = \dim \mathcal{L}(AC) = r(A)$.

Suppose now that $A \in L(X)$ is not of finite rank. For arbitrary $n \in N$ there exist projections $P,Q \in L(X)$ such that $r(AP) = r(QA) = n$. By the first part of the proof there exist $B,C \in L(X)$ such that $\dim \mathcal{L}(APC) = \dim \mathcal{L}(BQA) = n$. This proves the second assertion.□

We note that the results of Section 2 will show that if $A \in L(X)$ is such that $\dim \mathcal{L}(AB) < \infty$ (or $\dim \mathcal{L}(BA) < \infty$) for any $B \in L(X)$ then $A \in D(X)$.

Let $K(X)$ be the set of all <u>compact operators</u> in $L(X)$ (see [3], [4], [11]) . It is a closed two-sided ideal in $L(X)$ which contains $D(X)$. If X is a Hilbert space $K(X)$ is the closure of $D(X)$, but for general Banach spaces this may fail to be true.

Every $A \in K(X)$ has the following property:

$$(1)\quad \begin{cases} \sigma(A) \text{ has no accumulation point different from zero;} \\ \text{for any } \lambda \in \sigma(A),\ \lambda \neq 0 \text{, the corresponding spectral} \\ \text{projection of } A \text{ belongs to } D(X)\ . \end{cases}$$

Moreover, $K(X)$ being a two-sided ideal, the same property holds true also for BA and AB for any $B \in L(X)$. Now, consider the set $C(X)$ of all $A \in L(X)$ such that AB has the property (1) for any $B \in L(X)$. As a consequence of Theorem 3.1 this is a closed two-sided ideal in $L(X)$ equal to the set of all $A \in L(X)$ such that BA has the property (1) for any $B \in L(X)$.

Suppose now that X is a Hilbert space. Every hermitian $A \in L(X)$ with the property (1) is compact. Furthermore, A^*A is compact if and only if A is compact. It follows that in this case $C(X) = K(X)$.

In general, $C(X)$ can contain $K(X)$ properly. Although it is not necessary for the rest, I will mention here an interesting fact. Namely, there is another closed two-sided ideal in $L(X)$ which is between $K(X)$ and $C(X)$. This is the set $S(X)$ of all strictly singular operators on X [5]. $A \in L(X)$ is called strictly singular if for any infinite-dimensional subspace Y of X there does not exist $\alpha > 0$ such that $|Ay| \geq \alpha |y|$ for every $y \in Y$. Then $S(X)$ contains $K(X)$ and is contained in $C(X)$. Furthermore, it is known that for some Banach spaces $S(X) \neq K(X)$. It is an open problem whether there exists a Banach space X such that $S(X) \neq C(X)$.

$A \in L(X)$ is called a Fredholm operator if $\ker A$ is finite-dimensional and AX is of finite codimension in X (then necessarily AX is closed). Let $F(X)$ denote the set of all Fredholm operators on X. $F(X)$ can be defined also in another way [3]. Let $\pi : L(X) \to L(X)/K(X)$ be the canonical map. Then

(2) $$F(X) = \pi^{-1}(G(L(X)/K(X))).$$

In fact, by a result in [2] $K(X)$ can be substituted in (2) by any (not necessarily closed) two-sided ideal in $L(X)$ contained in $C(X)$ and containing $D(X)$. (2) shows that $F(X)$ is an open multiplicative subsemigroup of $L(X)$.

Now, if $A \in F(X)$, the index of A is an integer defined by

$$\mathrm{ind}_X(A) = \dim(\ker A) - \dim(X/AX).$$

Then the function $\mathrm{Ind}_X : F(X) \to Z$ has some remarkable properties. For example [3]:

(i) Ind_X is continuous, i.e. constant on every connected component of $F(X)$.

(ii) $\mathrm{Ind}_X(AB) = \mathrm{Ind}_X(A) + \mathrm{Ind}_X(B)$, $A, B \in F(X)$.

(iii) $\mathrm{Ind}_X(A+K) = \mathrm{Ind}_X(A) = \mathrm{Ind}_X(AT)$ for $A \in F(X)$, $K \in K(X)$, $T \in G(L(X))$.

(iv) For $A \in F(X)$ the following three properties are mutually equivalent:

(a) $\mathrm{Ind}_X(A) = 0$,

(b) $A = T+K$ for some $T \in G(L(X))$, $K \in K(X)$,

(c) $A = S+D$ for some $S \in G(L(X))$, $D \in D(X)$.

These properties make index a very powerful tool in operator theory [3].

Let $A \in F(X)$ and let $P,Q \in L(X)$ be projections with $\ker P = AX$ and $\ker A = QX$. Then P and Q are of finite rank and it is easy to see that $L(X)P = \{B \in L(X) : BA = 0\}$ and $QL(X) = \{B \in L(X) : AB = 0\}$. Furthermore, $\dim(\ker A) = r(Q)$ and $\dim(X/AX) = r(P)$. By Theorem 1.2 we obtain:

$$\dim(\ker A) = \sup\{\dim L(B) : BA = 0\}$$

$$\dim(X/AX) = \sup\{\dim L(B) : AB = 0\}.$$

This gives an idea for "algebraization" of the notion of index.

2. FINITE RANK ELEMENTS IN BANACH ALGEBRAS

In this section A will denote a complex Banach algebra with the identity 1. For $a \in A$ let $L(a)$ be the algebra generated by a:

$$L(a) = \operatorname{span}\{a, a^2, a^3, \ldots\}.$$

Set

$$d(a) = \dim L(a),$$

and for any subset S of A let

$$\delta(S) = \sup\{d(a) : a \in S\}$$

The key result, which allows "algebraization" of the notion of finite rank elements, is Theorem 2.2 stating that for certain subsets of A the finiteness of the function d implies its boundedness. Before stating it we have to establish an elementary fact about entire mappings between Banach spaces.

Let X and Y be Banach spaces. $f: X \to Y$ is called an <u>entire mapping</u> if it is continuous and if $\lambda \to f(x+\lambda y)$ is an entire Y-valued function on C for all $x, y \in X$.

LEMMA 2.1. Let $f: X \to Y$ be an entire mapping. If $f \neq 0$, then the set $N(F) = \{x \in X : f(x) = 0\}$ is closed and nowhere dense in X.

PROOF. $N(f)$ is closed by the continuity of f. Suppose that $N(f)$ is not nowhere dense, i.e. that there exists a non-empty open set U in X such that $f|U = 0$. Fix $x \in U$ and let $y \in X$. Then $\lambda \to f(x+\lambda y)$ is an entire Y-valued function on C vanishing on a neighbourhood of 0. Therefore, $f(x+\lambda y) = 0$ for all $\lambda \in C$. Since $y \in X$ was arbitrary, it follows $f = 0$. □

THEOREM 2.2. Let X be a Banach space and $F: X \to A$ an entire mapping. Suppose that $d(F(x)) < \infty$ for any $x \in X$. Then $\delta(F(X)) < \infty$ and the set $\{x \in X : d(F(x)) = \delta(F(X))\}$ is dense and open in X.

PROOF. For any non-negative integer n set $V_n = \{x \in X : d(F(x)) \leq n\}$ and $U_n = V_n \setminus V_{n-1}$. $X \setminus V_n$ is the set of all $x \in X$ such that $F(x), F(x)^2, \ldots, F(x)^{n+1}$ are linearly independent. By the continuity of F it follows that $X \setminus V_n$ is open. Thus, V_n is closed in X for every n. Since X is the union of V_n's, by Baire Category Theorem some V_k has non-empty interior. Let n be the smallest such k and denote by V the interior of V_n. Then $V \setminus V_{n-1}$ is open and non-empty and it is contained in U_n. Therefore, U_n has non-empty interior.

For $x \varepsilon U_n$ we can write

(1) $$F(x)^{n+1} = \sum_{j=1}^{n} a_j(x)\, F(x)^j, \quad a_j(x) \varepsilon C.$$

Fix $x_0 \varepsilon U_n$ and choose continuous linear functionals $f_1, \ldots, f_n$ on A such that

(2) $$f_i(F(x_o)^j) = \delta_{ij}, \quad i,j = 1, \ldots, n.$$

This is possible since $F(x_0), F(x_0)^2, \ldots, F(x_0)^n$ are linearly independent. Define the mappings $P_1, \ldots, P_n : X \to A'$ (= the space of continuous linear functionals on A), $P : X \to C$ and $Q : X \to A$ by

$$P_j(x) = \det \begin{bmatrix} f_1(F(x)) \ldots f_1(F(x)^{j-1})\ f_1\ \ f_1(F(x)^{j+1}) \ldots\ f_1(F(x)^n) \\ \cdot \\ \cdot \\ f_n(F(x))\ \ldots f_n(F(x)^{j-1}) f_n\ \ f_n(F(x)^{j+1}) \ldots\ f_n(F(x)^n) \end{bmatrix}$$

$$P(x) = P_j(x)(F(x)^j) = \det\, (f_i(F(x)^j)_{i,j=1}^{n})$$

$$Q(x) = P(x) \cdot F(x)^{n+1} - \sum_{j=1}^{n} P_j(x)(F(x)^{n+1}) F(x)^j .$$

Evidently, P and Q are entire mappings. From (1) we obtain $P_j(x)(F(x)^{n+1}) = a_j(x) \cdot P(x)$ for any $x \varepsilon U_n$, therefore

(3) $$Q(x) = 0, \qquad x \varepsilon U_n .$$

Since U_n has non-empty interior, (3) and Lemma 2.1 imply that $Q(x) = 0$ for every $x \varepsilon X$, i.e.

(4) $$P(x) \cdot F(x)^{n+1} = \sum_{j=1}^{n} P_j(x)(F(x)^{n+1}) F(x)^j, \qquad x \varepsilon X .$$

Now, (2) implies $P \neq 0$ so $N(P)$ is nowhere dense in X. By (4) $X \setminus V_n$ is contained in $N(P)$. Since $X \setminus V_n$ was shown to be open we conclude that $X \setminus V_n$ is empty, i.e. $X = V_n$. Therefore, $\delta(F(X)) = n < \infty$. Finally, $\{x \in X : d(F(x)) = n\} = U_n = X \setminus V_{n-1}$ and by the choice of n this set is open and dense in X. □

THEOREM 2.3. Let $a \in A$. The following six conditions are mutually equivalent:

(i) $d(ab) < \infty$ for every $b \in A$.

(ii) $d(ba) < \infty$ for every $b \in A$.

(iii) $d(bac) < \infty$ for every $b \in A$.

(iv) $\delta(aA) < \infty$.

(v) $\delta(Aa) < \infty$.

(vi) $\delta(AaA) < \infty$ (where $AaA = \{bac : b,c \in A\}$).

If this is the case, $\delta(aA) = \delta(Aa) = \delta(AaA)$.

PROOF. Theorem 2.2. applied to the case $X = A$ (resp. $X = A$, $X = A \times A$) and $F(x) = ax$ (resp. $F(x) = xa$, $F(x,y) = xay$) proves the equivalence (i)<=>(iv) (resp. (ii) <=>(v), (iii)<=>(vi)). Obviously, $\delta(aA) \leqq \delta(AaA)$ and $\delta(Aa) \leqq \delta(AaA)$. We have $(xay)^n = x(ayx)^{n-1}ay$, hence $\delta(AaA) \leqq \delta(aA)+1$. Similarly, $\delta(AaA) \leqq \delta(Aa)+1$. This shows the equivalence of (iv), (v) and (vi).

Suppose now that (i) - (vi) are satisfied and set $U = \{(x,y) \in A \times A : d(xay) = \delta(AaA)\}$. By Theorem 2.2 U is dense in $A \times A$. Since $G(A)$ is open in A, U contains a pair $(x,y) \in G(A) \times G(A)$. Now, $\delta(AaA) = d(xay) = d(x^{-1}(xay)x) = d(ayx) \leqq \delta(aA)$, and similarly $\delta(AaA) \leqq \delta(Aa)$. This proves the equality $\delta(aA) = \delta(Aa) = \delta(AaA)$. □

An element $a \in A$ satisfying the conditions of Theorem 2.3 will be called of finite rank. The number $\delta(AaA) = \delta(Aa) = \delta(aA)$ will be called the rank of a and denoted by $r(a)$. Let $D(A)$ be the set of all finite rank elements of A.

THEOREM 2.4. $D(A)$ is a two-sided ideal in A.

PROOF. (Sketch, see [12]). Similarly to the proof of Theorem 2.2 one can show that if $f: C \to A$ is a meromorphic function such that $d(f(\lambda)) < \infty$ for every $\lambda \in C \setminus P$ (P is the set of poles of f) and if $f(\lambda_0) \in G(A)$ for some $\lambda_0 \in C \setminus P$, then $\lambda \to f(\lambda)^{-1}$ is a meromorphic function on C. Furthermore, note that $d(a) < \infty$ is equivalent to the fact that $\lambda \to (\lambda - a)^{-1}$ is a meromorphic function on C. Now, let $a,b \in D(A)$. Then $\lambda - (a+b) = (\lambda - a)(1-(\lambda - a)^{-1}b)$ for $\lambda \notin \sigma(a)$. Furthermore, $f(\lambda) = 1-(\lambda - a)^{-1}b$ is meromorphic on C, $d(f(\lambda)) < \infty$ for $\lambda \notin \sigma(a)$ and $f(\lambda) \in G(A)$ for $|\lambda|$ large enough. Therefore, $\lambda \to f(\lambda)^{-1}$ is meromorphic on C and this implies that $\lambda \to (\lambda - (a+b))^{-1}$ is meromorphic on C. Hence, $d(a+b) < \infty$. Since the implications $c \in A$, $d \in D(A) \Rightarrow cd \in D(A)$ and $dc \in D(A)$ are obvious, the theorem follows. □

We note that Theorem 2.4 is proved in a completely algebraic way in [7] (Theorem 1 in X.14).

The following theorem shows that for a projection $p \in P(A)$ the property of being of finite rank depends only on the algebra pAp and $r(p)$ can be computed in this algebra.

THEOREM 2.5. Let $p \in P(A)$. Then p is of finite rank if and only if $\delta(pAp) < \infty$. In this case $r(p) = \delta(pAp)$.

PROOF. Obviously $r(p) \geq \delta(pAp)$. Suppose that $\delta(pAp) = n < \infty$. Applying Theorem 2.2 to the case $X = A$ and $F(x) = pxp$ we see that there exists $a \in A$ such that $d(pap) = n$ and $pap \in G(pAp)$. Let $b \in pAp$ be the inverse of pap in pAp, i.e. such that $bpap = papb = p$. We have for $j \in N$, $(ap)^j = p(ap)^j + (1-p)(ap)^j = (pap)^j + (1-p)a(pap)^{j-1} = (1+(1-p)ab)(pap)^j$. Hence, for $\lambda_1, \dots, \lambda_k \in C$

$$\sum_{j=1}^{k} \lambda_j (ap)^j = 0 \iff \sum_{j=1}^{k} \lambda_j (pap)^j = 0 .$$

Therefore, $d(ap) = n$ for every $a \in A$ such that $d(pap) = n$ and $pap \in G(pAp)$. By Theorem 2.2 the set of all such elements $a \in A$ is open in A and non-empty. Applying the same theorem to the case $X = A$, $F(x) = xp$, we conclude that $\delta(Ap) = n$. □

An easy consequence is the following:

COROLLARY 2.6. Let $p,q \in D(A) \cap P(A)$ be mutually orthogonal (i.e. $pq = qp = 0$). Then $r(p+q) = r(p) + r(q)$. □

3. FREDHOLM ELEMENTS AND INDEX IN BANACH ALGEBRAS

In this section A continues to denote a complex Banach algebra with the identity 1.

Let $R(A)$ be the set of all Riesz elements in A, i.e. the set of all $a \in A$ with the property : $\sigma(a)$ has no accumulation point different from zero, and for every $\lambda \in \sigma(a)$, $\lambda \neq 0$, the corresponding spectral projection is in $D(A)$.

An element $a \in A$ will be called compact if $ab \in R(A)$ for any $b \in A$. Let $C(A)$ denote the set of all compact elements in A.

The proof of the following theorem is rather long and will be omitted (see [12],[10]).

THEOREM 3 1. $C(A)$ is a closed two-sided ideal in A. Especially, $C(A)$ is the set of all $a \in A$ such that $ba \in R(A)$ for any $b \in A$.

Now, let $\pi : A \to A/C(A)$ be the canonical map and set

$$F(A) = \pi^{-1}(G(A/C(A))) .$$

The elements of $F(A)$ will be called Fredholm. $F(A)$ is an open multiplicative subsemigroup of A.

For $a \in A$ define the following closed left (resp. right) ideal in A :

$$L(a) = \{ x \in A : xa = 0 \} ,$$

$$R(a) = \{ x \in A : ax = 0 \} .$$

PROPOSITION 3.2. For $a \in F(A)$ the ideals $L(a)$ and $R(a)$ are contained in $D(A)$.

PROOF. By the definition of $F(A)$ there exist $h,k \in C(A)$ and $b \in A$ such that $ab = 1-k$, $ba = 1-h$. Obviously, $L(a) \subseteq L(1-k)$ and $R(a) \subseteq R(1-h)$. Thus, the proof reduces to the case $a = 1-k$, $k \in C(A)$.

Let $p \in D(A)$ be the spectral projection of k corresponding to 1. Set $n = pk-p$, $h = k-pk$. Then $k = p+n+h$, $pn = np = n$, $ph = hp = 0$. Let $b \in R(1-k)$. Then $kb = b$, hence $pb + nb + hb = b$. Thus, $pb = pb + nb$, hence $nb = 0$ and so $b = pb + hb$, i.e. $pb = (1-h)b$. Then $(1-h)(1-p)b = (1-p)(1-h)b = (1-p)pb = 0$. Spectral

Mapping Theorem implies $1 \notin \sigma(h)$ so we conclude that $(1-p)b = 0$. Therefore, $b = pb$, i.e. $b \in pA$. This shows that $R(1-k) \subseteq pA$. Since $p \in D(A)$, it follows that $R(1-k)$ is contained in $D(A)$. The proof for $L(1-k)$ is quite analogous. □

Now, since $L(a)$ and $R(a)$ are closed, as a consequence of Theorem 2.2 we have that $\delta(L(a)) < \infty$ and $\delta(R(a)) < \infty$. Motivated by the end of Section 1 we define the function $ind: F(A) \to Z$ by

$$ind(a) = \delta(R(a)) - \delta(L(a))$$

If $A = L(X)$ (X a Banach space) then $F(A) = F(X)$ and $\delta(R(a)) = \dim(\ker a)$, $\delta(L(a)) = \dim(X/aX)$. Therefore, the function ind generalizes the classical index.

CONJECTURE 3.3. The function $ind: F(A) \to Z$ has the following properties:

(i) ind is continuous, i.e. constant on every connected component of $F(A)$.

(ii) $ind(ab) = ind(a) + ind(b)$, $a,b \in F(A)$.

(iii) $ind(ab) = ind(ba) = ind(a+k) = ind(a)$ for $a \in F(A)$, $b \in G(A)$, $k \in C(A)$.

The main problem in proving this conjecture seems to be (ii). Some evidence about the truth of this conjecture is the case of semi-simple Banach algebras which is considered in the next section.

4. SEMISIMPLE BANACH ALGEBRAS

For an algebra A we shall denote its radical by $rad(A)$ (= the two-sided ideal of all $a \in A$ such that ab is quasinilpotent for every $b \in A$) ; $rad(A)$ contains any ideal in A consisting of quasinilpotent elements.

An algebra A is called algebraic if $d(a) < \infty$ for every $a \in A$. In this case $a \in A$ is quasinilpotent if and only if it is nilpotent. Especially, $rad(A)$ consists of nilpotent elements. If A has an identity then every $a \in A$ has a minimal polynomial μ and the spectrum of a is exactly the set of all zeros of μ.

Our first goal in this section is to establish the structure of semisimple (i.e. with zero radical) algebraic Banach algebras. For any $n \in N$ $M(n)$ will denote the full matrix algebra of all n by n complex matrices. Recall that a finite-dimensional algebra is semisimple if and only if it is isomorphic to a direct product of full matrix algebras; especially, it has an identity (cf. [6] ,1.4).

THEOREM 4.1. Let A be a complex semisimple algebraic Banach algebra. Then A is finite-dimensional.

PROOF. We can suppose that A has an identity 1 . In fact, if A does not have an identity, the Banach algebra B obtained from A by adjoining the identity is semisimple and algebraic and $\dim A < \infty$ follows from $\dim B < \infty$.

Suppose first that $P(A) = \{0,1\}$. Let $a \in A$ and $\lambda \in \sigma(a)$. The spectral projection of a corresponding to λ is different from 0 hence it is equal to 1. By Spectral Mapping Theorem $\sigma(a-\lambda) = \{0\}$, therefore $n = a-\lambda$ is nilpotent. Thus, any $a \in A$ has the form $a = \lambda + n$, n nilpotent. It follows that the set of all nilpotent elements in A is a two-sided ideal, hence it is contained in $rad(A)$. But $rad(A) = \{0\}$ and we conclude that $A = C$.

By Theorem 2.2 the degrees of the minimal polynomials of the elements of A is bounded, say by $n \in N$. Then the cardinality of a family of non-zero mutually orthogonal projections is also bounded by n . Indeed, if $p_1,\dots,p_k \in P(A)$ are mutually orthogonal then the spectrum of $a = p_1+2p_2+\dots+kp_k$ has cardinality k ; but then $k \leq n$ because the degree of the minimal polynomial of a is at most n .

Let $\{p_1,\dots,p_m\}$ be a maximal family of non-zero mutually orthogonal projections in A. Then $p_1+\dots+p_m = 1$; indeed, otherwise $1 - p_1-\dots-p_m$ would be a non-zero projection orthogonal to $p_1,\dots,p_m$. Now, p_jAp_j is a semisimple algebraic Banach algebra and from the maximality of the family $\{p_1,\dots,p_m\}$ it follows that $P(p_jAp_j) = \{0,p_j\}$. By the second paragraph in the proof we conclude that $p_jAp_j = Cp_j$. We prove now that $\dim p_iAp_j \leqq 1$ for any $i \neq j$. Interchanging A by qAq , $q = p_i+p_j$, the proof is reduced to the case $p_i + p_j = 1$. Now, $p_iAp_i = Cp_i$ and $p_jAp_j = Cp_j$ imply that there are linear forms f,g on A such that

$$p_iap_i = f(a)p_i \ , \quad p_jap_j = g(a)p_j \ , \quad a \in A \ .$$

The equality $ab = ap_ib+ap_jb$ easily implies that $f(ab) = f(a)f(b) + f(ap_jb)$, $g(ab) = g(a)g(b) + g(ap_ib)$, $p_iabp_j = f(a)p_ibp_j + g(b)p_iap_j$, $p_jabp_i = g(a)p_jbp_i + f(b)p_jap_i$. From there and from the associativity it follows that $f(ap_jb)p_icp_j = g(bp_ic)p_iap_j$ for any $a,b,c \in A$ .Thus, if $\dim p_iAp_j > 1$, for every $p_iap_j \neq 0$ there is a linearly independent p_icp_j . Therefore $f(ap_jb) = 0$ for any $a,b \in A$. Now, for $a,b \in A$ we have $p_iap_jb = p_iap_jbp_i + p_iap_jbp_j = f(ap_jb)p_i + p_iap_jbp_j = p_iap_jbp_j$. This shows that p_iAp_j is a right ideal in A . For any element $a \in p_iAp_j$ we have $a^2 = 0$. Thus $p_iAp_j \subsetneq \mathrm{rad}(A) = \{0\}$, a contradiction.

Therefore, $\dim p_iAp_j \leqq 1$ for any $i,j = 1,\dots,m$. Since A is the sum of p_iAp_j , it follows that $\dim A \leqq m^2$. □

In the rest of this section A will denote a semisimple complex Bacah algebra with the identity. We denote $P_d(A) = P(A) \cap D(A)$.

PROPOSITION 4.2. Let I be a closed left (resp. right) ideal in A contained in $D(A)$. Then I has a right (resp. left) identity $p \in P_d(A)$, i.e. $I = Ap$ (resp. $I = pA$) . If $q \in I \cap P_d(A)$, p can be chosen so that $pq = qp = q$.

PROOF. $B = I/\mathrm{rad}(I)$ is a semisimple algebraic Banach algebra. Theorem 4.X. 11 in [7] and the remark following it imply that I contains a subalgebra D such that $I = D \dotplus \mathrm{rad}(I)$ and $q \in D$. Then D is isomorphic to B , thus it possesses an identity p. Then $pq=qp=q$. p is the identity of A. Indeed, otherwise for some $a \in I, 0\neq r=ap-a \in I$. Let $b \in A$ be such that br is not nilpotent; such b exists since A is semisimple. Now, if $br = c+d$, $c \in \mathrm{rad}(I)$, $d \in D$, we have

$d = dp = (br-c)p = -pc$, since $rp = 0$. Thus $d \in \mathrm{rad}(I) \cap D = \{0\}$, i.e. $br \in \mathrm{rad}(I)$. But this is impossible because $\mathrm{rad}(I)$ consists of nilpotent elements.□

Now, for $a \in F(A)$ the ideals $L(a)$ and $R(a)$ are closed and contained in $D(A)$, therefore there exist $p,q \in P_d(A)$ such that $L(a) = Ap$ and $R(a) = qA$. Furthermore, $r(p) = \delta(Ap) = \delta(L(a))$ and similarly $r(q) = \delta(R(a))$. Thus,

$$\mathrm{ind}\,(a) = r(q) - r(p) \ .$$

We will use it to prove the properties of the function ind. For this we need some further properties of the rank function.

The following proposition will allow us to reduce many proofs to the case of matrix algebras.

PROPOSITION 4.3. Let $a_1,\dots,a_n \in D(A)$. There exists $p \in P_d(A)$ such that $a_1,\dots,a_n \in pAp$.

PROOF. $I = Aa_1+\dots+Aa_n$ is a left ideal in A contained in $D(A)$. By Theorem 2.2 applied to $X = A\times\dots\times A$ (n times) and $F(x_1,\dots,x_n) = x_1a_1 + \dots + x_na_n$ it follows that $\delta(I) < \infty$. Now $\delta(\bar{I}) = \delta(I)$, hence $\bar{I}$ is contained in $D(A)$. By Proposition 4.2 there exists $q \in P_d(A)$ such that $\bar{I} = Aq$. Then $a_iq = a_i$ for $i = 1,\dots,n$.

The same reasoning applied to the right ideal $a_1A+\dots+a_nA+qA$ shows that there exists $p \in P_d(A)$ such that $pa_i = a_i$, $i = 1,\dots,n$, and $pq = q$. By the second assertion in Proposition 4.2 p can be chosen so that $qp = q$. It follows $pa_i=a_ip=a_i$, $i = 1,\dots,n$, i.e. $a_1,\dots,a_n \in pAp$. □

PROPOSITION 4.4. An ideal I in A contained in $D(A)$ is closed if any only if $\delta(I) < \infty$.

PROOF. If I is closed Theorem 2.2 $(X = I,\ F(x) = x)$ implies that $\delta(I) < \infty$. Suppose that I is a left ideal such that $\delta(I) < \infty$. Then $\delta(\bar{I}) = \delta(I) < \infty$, thus $\bar{I} \subseteq D(A)$. By Proposition 4.2 $\bar{I} = Ap$ for some $p \in P_d(A)$. By approximating p with elements of I and then taking their spectral projections corresponding to non-zero spectrum one shows that p can be approximated by projections in I . Take $q \in I \cap P_d(A)$ such that $|p-q| < 1$. Then p and q are $G(A)$-conjugated, hence $r(p) = r(q)$. This, together with $qp = q$ implies $pq = p$; to prove this one can use Proposition 4.3 to reduce it to an easy exercise in linear algebra. Therefore, $p \in I$ and $\bar{I} = I$. □

THEOREM 4.5. Let $a \in F(A)$ and $p \in P_d(A)$. Then $L(A) = Ap$ if and only if $aA = (1-p)A$. Similarly, for $q \in P_d(A)$, $R(a) = qA$ if and only if $Aa = A(1-q)$. Especially, the ideals aA and Aa are closed.

PROOF. Let $p \in P_d(A)$ be such that $aA = (1-p)A$. Then $1-p=ab$ for some $b \in A$, hence $L(a) \subseteq Ap$. Similarly, we have $a = (1-p)c$ for some $c \in A$ and this implies $Ap \subseteq L(a)$. Thus, $aA = (1-p)A \Rightarrow L(a) = Ap$.

Now, let $p \in P_d(A)$ be such that $L(a) = Ap$. Choose $b \in A$ so that $ab = 1-k$, $k \in C(A)$. Then $(1-k)A \subseteq aA$. Let p_0 be the spectral projection of k corresponding to 1 . Then $(1-p_0)A \subseteq (1-k)A$. Thus $aA = (1-p_0)A \dotplus p_0A \cap aA$. Since $\delta(p_0A \cap aA) \leq \delta(p_0A) = r(p_0) < \infty$, Propositions 4.3 and 4.2 imply that $p_0A \cap aA = p_1A$ for some $p_1 \in P_d(A)$. Interchanging p_1 by p_1p_0 if necessary, we can suppose that $p_1p_0 = p_0p_1 = p_1$. Then $q = p_0-p_1 \in P_d(A)$ and we have $aA = (1-q)A$. The first part of the proof implies $L(a) = Aq$. Thus $Ap = Aq \Rightarrow pq = p$ and $qp = q \Rightarrow (1-p)(1-q) = 1-q$ and $(1-q)(1-p) = 1-p \Rightarrow (1-p)A = (1-q)A$. Therefore $aA = (1-p)A$. □

LEMMA 4.5. Let $a \in A$ and $p \in P_d(A)$ be such that $R(a) \cap pA = \{0\}$. Then $r(ap) = r(p)$.

PROOF. Let $q \in P_d(A)$ be such that $ap, p \in qAq$. Then $R(a) \cap pA = \{0\}$ implies that qaq is invertible in qAq, hence $r(ap) = r(qapq) = r(qaqp) = r(p)$. □

LEMMA 4.6. Let $p,q \in P_d(A)$ and suppose that either $A = pA \dotplus (1-q)A$ or $A = Ap \dotplus A(1-q)$ or $pA \cap (1-q)A = A(1-p) \cap Aq = \{0\}$. Then $r(p) = r(q)$.

PROOF. Taking $P \in P_d(A)$ such that $p,q \in PAP$ the proof again reduces to linear algebra. □

THEOREM 4.7. Conjecture 3.3. is true for semisimple Banach algebras.

PROOF. Let $a,b \in F(A)$ and choose $e,f,g,h,j,k,p,q \in P_d(A)$ so that $R(a) = eA$, $L(a) = Af$, $R(b) = gA$, $L(b) = Ah$, $R(ba) = jA$, $L(ba) = Ak$, $R(b) \cap aA = pA$, $L(a) \cap Ab = Aq$.

We prove first that

(1) $$r(j) = r(e) + r(p) .$$

Let $u \in P_d(A)$ be such that $(1-e)A \cap R(1-g)a = uA$. We can suppose that $ue = eu = 0$ (if not, interchange u by $u(1-e)$). It follows easily that $auA = aA \cap gA = pA$. Now, $uA \subseteq (1-e)A$ implies $R(a) \cap uA = \{0\}$, and by Lemma 4.5 we obtain $r(p) = r(u)$. Easy calculations show that $R(ba) = eA \dot{+} uA = (e+u)A$ and $e+u \in P_d(A)$ since $eu = ue = 0$. Using Corollary 2.6 we have $r(j) = \delta(R(ba)) = \delta((e+u)A) = r(e+u) = r(e) + r(u) = r(e) + r(p)$.

Quite similarly one proves that

(2) $$r(k) = r(h) + r(q) .$$

Finally, we prove that

(3) $$r(g) - r(p) = r(f) - r(q) .$$

To do this we can suppose that $gp = pg = p$ and $fq = qf = q$ (if not, interchange p by pg and q by fq). By Corollary 2.6 we have to prove that $r(p_0) = r(q_0)$, where $p_0 = g-p$, $q_0 = f-q \in P_d(A)$.

Let $x \in p_0A \cap (1-q_0)A$. $p_0A \subseteq gA$ implies $gx = x$ and $Aq \subseteq A(1-g)$ implies $qg = 0$. Thus, $qx = 0$. Since $x \in (1-q_0)A$ it follows $fx = 0$, therefore $x \in p_0A \cap (1-f)A \subseteq gA \cap (1-f)A$ and by Theorem 4.5. $x \in R(b) \cap aA = pA$. This implies $x = px = pp_0x = 0$. Therefore, $p_0A \cap (1-q_0)A = \{0\}$. Quite similarly one proves that $A(1-p_0) \cap Aq_0 = \{0\}$. Now, by Lemma 4.6 we obtain $r(p_o) = r(q_0)$.

The equalities (1), (2) and (3) give

$$\mathrm{ind}(ba)=r(j)-r(k)=r(e)-r(f)+r(g)-r(h)= \mathrm{ind}\ (b)+ \mathrm{ind}\ (a) ,$$

i.e. (ii) of Conjecture 3.3 is true.

Obviously, ind $(b) = 0$ for any $b \in G(A)$, so (ii) implies ind (ab) = ind (ba) = ind (a) , $a \in F(A)$, $b \in G(A)$.

Let $k \in K(A)$. If $1 \notin \sigma(k)$ then $1-k \in G(A)$ and ind $(1-k) = 0$. Suppose $1 \in \sigma(k)$. Let p be the spectral projection of k corresponding to 1 and set $n = pk - p$. Then $n \in D(A)$ is nilpotent. Let $e,f \in P_d(A)$ be such that $L(1-k) = Ae$, $R(1-k) = fA$. Choose $P \in P_d(A)$ so that $n,p,e,f \in PAP$. One easily obtains $pn = np = n$, $Ae = L(n) \cap Ap$, $fA = R(n) \cap pA$. It follows that $PAPe = L(n) \cap PAPp$,

$fPAP = R(n) \cap pPAP$. Since PAP is a direct product of full matrix algebras, an exercise in linear algebra shows that it follows $r(e) = r(f)$ (to prove this use Jordan basis for n in which p diagonalizes). Therefore $\text{ind}\,(1-k) = 0$ for any $k \in K(A)$.

Now, take $a \in F(A)$ and $k \in K(A)$. Choose $b \in F(A)$ so that $ba = 1 + h$, $h \in K(A)$. Then $h+bk \in K(A)$ and by (ii) and the preceding paragraph it follows that $\text{ind}\,(b) + \text{ind}\,(a) = \text{ind}\,(ba) = \text{ind}(1+h) = 0$ and $\text{ind}\,(b) + \text{ind}\,(a+k) = \text{ind}\,(ba + bk) = \text{ind}\,(1+(h+bk)) = 0$. Thus, $\text{ind}\,(a+k) = \text{ind}\,(a)$. This proves (iii) of Conjecture 3.3.

Finally, let us prove (i). Let $a \in F(A)$, $L(a) = Ap$, $R(a) = qA$, $p,q \in P_d(A)$. $X = (1-q)A \times pA$ is a Banach space with the norm $\|(v,w)\| = |v| + |w|$. For $x \in A$ let $S_x \in L(X,A)$ be defined by $S_x(v,w) = xv + w$. We assert that S_a is a bijection of X onto A. Indeed, $S_a(v,w) = 0$ implies $av = -w \in aA \cap pA = (1-p)A \cap pA = \{0\}$, because of Theorem 4.5. Therefore, $w = 0 = av \Rightarrow v \in R(a) \cap (1-q)A = \{0\} \Rightarrow v = 0$. Hence, S_a is an injection. Let $c \in A$. Then $(1-p)c \in (1-p)A = aA$, hence $(1-p)c = ab$ for some $b \in A$. Now, $S_a((1-q)b,pc) = a(1-q)b+pc = ab+pc = c$. Hence, S_a is also a surjection. Open Mapping Theorem implies that S_a is topological isomorphism of Banach spaces. Now, $x \to S_x$ is continuous from A to $L(X,A)$, hence S_x is a topological isomorphism for all x in a neighbourhood of a. (i) will be proved if we show that $\text{ind}\,(a) = \text{ind}\,(b)$ if $b \in F(A)$ is such that S_b is a bijection. Set $c = b(1-q)$. By (iii) we have $\text{ind}\,(c) = \text{ind}\,(b-bq) = \text{ind}\,(b)$. For $x \in R(c)$ we have $S_b(1-q)x,0) = 0$, hence $(1-q)x = 0$. Thus $R(c) = R(1-q) = qA$. Choose $s \in P_d(A)$ so that $L(c) = As$. By Theorem 4.5 we have $(1-s)A = cA = b(1-q)A$. Since S_b is an isomorphism, this gives $A = (1-s)A \dotplus pA$. By Lemma 4.6 we obtain $r(p) = r(s)$. Therefore, $\text{ind}\,(b) = \text{ind}\,(c) = r(q)-r(s) = r(q) - r(p) = \text{ind}\,(a)$. □

We note that Conjecture 3.3 was shown to be true even for larger class of semiprime Banach algebras [2].

5. REFINEMENT OF THE DEFINITION OF INDEX

One would like that the important property (iv) of Section 1 is satisfied by the index function. In our general framework of Banach algebras this property reads:

(1) For $a \in A$ the following three properties are mutually equivalent:
- (a) $\mathrm{ind}(a) = 0$,
- (b) for some $u \in G(A)$ and $k \in C(A)$, $a = u+k$,
- (c) for some $v \in G(A)$ and $d \in D(A)$, $a = v+d$.

It is not difficult to see that at least for semisimple algebras (even for larger class of semiprime algebras) (b) is equivalent to (c) and implies (a) . But (a) does not imply (b) even for semisimple algebras. To see this consider an example. Let X be a Banach space, Y and Z closed infinite-dimensional subspaces such that $X = Y \dot{+} Z$ and $A = \{T \in L(X) : TY \subseteq Y , TZ \subseteq Z \}$. Then $F(A)$ can be identified with $F(Y) \dot{+} F(Z)$ and for $T \in F(Y)$, $S \in F(Z)$, $\mathrm{ind}(T \dot{+} S) = \mathrm{Ind}_Y(T) + \mathrm{Ind}_Z(S)$. Thus, if we take T and S such that $\mathrm{Ind}_Y(T) = -\mathrm{Ind}_Z(S) \neq 0$, then $\mathrm{ind}(T \dot{+} S) = 0$. But it is not possible to write $T \dot{+} S = U+K$ with $U \in G(A) = G(L(Y)) \dot{+} G(L(Z))$ and $K \in C(A) = C(Y) \dot{+} C(Z)$. However, this elementary example gives us an idea how one should try to redefine the notion of index so that (1) would be satisfied. Namely, if we define $\mathrm{ind}: F(A) \to Z^2$ by $\mathrm{ind}(T \dot{+} S) = (\mathrm{Ind}_Y(T), \mathrm{Ind}_Z(S))$, then (1) is satisfied.

Let us make this definition more general. Suppose A is a complex <u>semiprime</u> Banach algebra with the identity (semiprime means that there are not ideals $I \neq \{0\}$ in A such that $I^2 = \{0\}$). In this case it is known that A has a lot of projections ; in fact, it is shown in [2] that for any $a \in F(A)$ there exist $p,q \in P_d(A)$ such that $L(a) = Ap$, $R(a) = qA$. Let $P_m(A)$ be the set of all minimal elements of $P_d(A) \setminus \{0\}$ with respect to the natural order relation

$$p \leq q \iff pq = qp = p .$$

$G(A)$-conjugacy gives us an equivalence relation in $P_m(A)$:

$$p \sim q \iff p = uqu^{-1} \quad \text{for some } u \in G(A) .$$

Let $\Gamma = P_m(A)/\sim$. For each $\gamma \in \Gamma$ define $r_\gamma : P_d(A) \to Z$ in the following way: if $p \in P_d(A)$ with $r(p) = n$, then $p = p_1 + \ldots + p_n$ with $p_i \in P_m(A)$ and $p_i p_j = 0$ for $i \neq j$; set

$$r_\gamma (p) = \mathrm{Card}\ \{ i : 1 \leq i \leq n , p_i \in \gamma \} .$$

Now, for $a \in F(A)$ take $p,q \in P_d(A)$ such that $L(a) = Ap$, and $R(a) = qA$ and for $\gamma \in \Gamma$ set

(2) $$\mathrm{Ind}_\gamma (a) = r_\gamma (q) - r_\gamma (p) .$$

Define $\mathrm{Ind} : F(A) \to Z^{(\Gamma)}$ by

(3) $$\mathrm{Ind}\ (a) = (\mathrm{Ind}_\gamma (a))_{\gamma \in \Gamma} \quad ;$$

here $Z^{(\Gamma)}$ denotes the direct sum of Γ copies of Z , or the group of all finitely supported functions $\Gamma \to Z$.

We note that for $A = L(X)$ Γ has only one element and in the case of the example in this section Γ has two elements. There exist algebras which are not direct products of two-sided ideals but still $\mathrm{Card}\ \Gamma > 1$ (see [9]).

Properties of this index function are investigated in [9] in the case when A is semisimple. It is shown that Conjecture 3.3 is true in this case and, moreover, that (1) is satisfied.

CONJECTURE 5.1. If A is a complex semiprime Banach algebra with the identity Conjecture 3.3 and the property (1) are satisfied by the function Ind defined by (2) and (3).

This conjecture has some interesting consequences. For example, as it is shown in [9] if $\pi : A \to A/C(A)$ is the quotient map $\pi(G(A))$ is a normal open subgroup of $G(A/C(A))$ and the quotient group $G(A/C(A))/\ \pi(G(A))$ is Abelian and without torsion. In fact it is easy to see that Ind induces an injective homomorphism of this group into $Z^{(\Gamma)}$. This holds true also for A/I instead of $A/C(A)$, where I is any closed two-sided ideal in A contained in $C(A)$ and containing $D(A)$.

Concluding, we state one more unsolved problem.

PROBLEM 5.2. How to modify the definition of the index function for general Banach algebras so that Conjecture 3.3 together with the property (1) would be satisfied?

REFERENCES

[1] B.A.Barnes, A generalized Fredholm theory for certain maps in the regular representations of an algebra, Canad. J.Math. 20 (1968), 495-504.

[2] B.A.Barnes, The Fredholm elements of a ring, Canad.J.Math. 21 (1969), 84-95.

[3] R.G.Douglas, Banach algebra techniques in operator theory, Academic Press, New York and London 1972.

[4] N.Dunford, J.T.Schwartz, Linear operators, New York, I 1958, II 1963.

[5] S.Goldberg, Unbounded linear operators, McGraw-Hill, New York, 1966.

[6] I.N.Herstein, Noncommutative rings, J.Wiley, New York, 1968.

[7] N.Jacobson, Structure of rings, AMS Coll.Publ. 36, Providence, 1956.

[8] H.Kraljević, K.Veselić, On algebraic and spectrally finite Banach algebras, Glasnik Mat. 11(31) (1976), 291-318.

[9] H.Kraljević, S.Suljagić, K.Veselić, Index in semisimple Banach algebras, Glasnik Mat. 17 (37) (1982).

[10] L.D.Pearlman, Riesz points of the spectrum of an element in a semisimple Banach algebra, Trans.Amer.Math.Soc. 193 (1974), 303-328.

[11] A.E.Taylor, Introduction to functional analysis, J.Wiley, New York, 1958.

[12] K.Veselić, On essential spectra in Banach algebras, Glasnik Mat. 10 (30) (1975), 295-309.

FIXED POINT THEOREMS IN NOT NECESSARILY LOCALLY CONVEX TOPOLOGICAL VECTOR SPACES

Olga Hadžić

INTRODUCTION

In the last twenty years there is an increasing interest in the fixed point theory in not necessarily locally convex topological vector spaces. If X is a topological space and $K \subseteq X$ we say that the set K has <u>the fixed point property</u> if for every continuous mapping $f: K \to K$, $\mathrm{Fix}(f) = \{x \mid x \in K, x = fx\} \neq \emptyset$. Every compact and convex subset of a normed space has the fixed point property (Schauder's fixed point theorem) and more generaly every compact and convex subset of a locally convex topological vector space has the fixed point property (Tychonoff's fixed point theorem). Now, an open question is : Does every compact and convex subset of an arbitrary topological vector space has the fixed point property? Some interesting results about the topological structure of a class of non-locally convex topological vector spaces were obtained by Dobrowolski and Torunczyk in [2].

In [23] V.Klee introduced the notion of admissible subset of a topological vector space and for admissible topological vector spaces some fixed point theorems are proved by Klee [24], Landsberg [29], Landsberg-Riedrich [30], Riedrich [38], Schulz [41], and Hahn and Pötter [18]. If E is a Hausdorff topological vector space and J is a compact retract of E which admits arbitrarily small continuous displacement into a finite dimensional subspace of E then J has the fixed point property [24]. An interesting fixed point theorem, which is in fact a generalization of Tychonoff's fixed point theorem, is proved by Rzepecki in [39]. His result generalizes Zima's result from [43]. The Leray-Schauder theory without local convexity is given by Klee in [24]. Some fixed point theorems for not necessarily locally convex topological vector space were proved by C.Krauthausen ([26],[27]). He proved the admissibility of some non locally convex topological vector space [26] and introduced the notion of locally convex subset of a topological vector space [27]. Using the theory of topological semifield the author has proved some fixed point theorems in not necessarily locally convex topological vector space ([6],[7]). For mutivalued mappings a fixed point theorem in not necessarily locally convex topological vector space were proved by Hahn [16], Kaballo [21] and the author [9].

Here we shall give a survey of some of the above results and some new results.

1. FIXED POINT THEOREMS FOR SINGLE VALUED MAPPINGS IN NOT NECESSARILY LOCALLY CONVEX TOPOLOGICAL VECTOR SPACE

In this paper all topological vector spaces are Hausdorff. First, we shall give the definition of admissibility ([18] and [24]).

DEFINITION 1. Let E be a topological vector space and $Z \subseteq E$. The set Z is admissible if and only if for every compact subset $K \subseteq Z$ and for every neighbourhood V of zero in E there exists a compact mapping $h: K \to Z$ such that $\dim(\text{span } h(K)) < \infty$ and $x - hx \in V$, for every $x \in K$. If $Z = E$ then E is an admissible topological vector space.

The admissibility of the spaces $L^p(0,1)$ $(0 < p < 1)$ and $S(0,1)$ is proved by Riedrich in [36] and [37], of the Hardy spaces H^p $(0 < p < 1)$ by Krauthausen in [26], of the spaces $(S(X,A,m),d)$ and $(SH_\phi, \|\cdot\|_\phi)$ by Krauthausen in [27] and some other spaces in [7] and [20].

Some sufficient conditions for the admissibility of a subset of a topological vector space are given in [23] by Klee, in [26] and [27] by Krauthausen and in [7] by the author. V.Klee proved that every ultrabarrelled Hausdorff topological vector space which admits a Schauder basis is admissible, and that every admissible complete metric topological vector space is an extension space for the class of all compact Hausdorff spaces.

It is known that every convex subset of a locally convex topolo-ical vector space is admissible. An open question is: Is every convex subset of an arbitrary topological vector space admissible? An example of a nonconvex, compact, nonadmissible subset of ℓ_2 is :

$$\{(t^n)_{n \in N} \mid 0 \leq t \leq \tfrac{1}{2}\}.$$

For admissible subsets of topological vector spaces some interesting fixed point theorems are proved in [27] and the following generalization of Tychonoff's fixed point theorem is proved: Let E be a topological vector space and K an admissible, closed and convex subset of E. If $f: K \to K$ is a compact mapping then $\mathrm{Fix}(f) \neq \emptyset$. In [18] an application on the integral equation

$$y(s) = x(s) - \int_0^1 K(s,t) \frac{x(t)}{1+x(t)} dt, \quad s \in [0,1]$$

is given $(y \in S(0,1),\ K \in C_{[0,1]\times[0,1]})$.

In fact, in [18] the following main result is obtained.

THEOREM 1. Let E be an admissible topological vector space, W a closed neighbourhood of E and f a compact mapping of W into E so that $x \in \partial W$, $fx = \alpha x \Rightarrow \alpha \leq 1$. Then $\mathrm{Fix}(f) \neq \emptyset$.

PROOF. From $\mathrm{Fix}(f) = \emptyset$ and the admissibility of E it follows that for every neighbourhood of zero V in E there exists $h_V : W \to E$ such that $h_V(x) - x \in V$, for every $x \in W$, $\dim(\mathrm{span}\, h_V(W)) < \infty$, and h_V is a compact mapping with the property that $\mathrm{Fix}(h_V) = \emptyset$. Let $W_V = W \cap E_V$ (E_V is a finite dimensional subspace of E such that $h_V(W) \subseteq E_V$) and $\bar{h}_V = h_V | W_V$. Then there exists $x_V \in \partial_V W_V \subseteq \partial W$ (∂_V in the induced topology of E_V) and $\lambda_V \in (0,1)$ so that $x_V = \lambda_V \bar{h}_V(x_V)$. Since V is an arbitrary neighbourhood of zero in E it follows that there exists $\lambda_o \in (0,1)$ and $x_o \in \partial W$ so that $x_o = \lambda_o f(x_o)$. Since $\frac{1}{\lambda_o} > 1$ we obtain a contradiction and so $\mathrm{Fix}(f) \neq \emptyset$.

Similarly, the following fixed point theorem can be proved [18]:

Let E be a topological vector space, W a closed neighbourhood of E and K an admissible, closed and convex subset of E such that $0 \in K$. Further, let f be a compact mapping from $W \cap K$ into K so that $x \in K \cap \partial W$, $fx = \alpha x$ implies $\alpha \leq 1$. Then $\mathrm{Fix}(f) \neq \emptyset$.

In [27] C.Krauthausen has introduced the following definition:

DEFINITION 2. Let X be a topological vector space, $\mathcal{U}$ the family of all neighbourhoods of zero in X and $K \subseteq X$ a convex subset. The topology of K is locally convex (or K is locally convex) if and only if for every $V \in \mathcal{U}$ and every $x \in K$ there exists $U \in \mathcal{U}$ so that:

$$\mathrm{co}((x+U) \cap K) \subseteq x+V .$$

It is proved in [27] that a locally convex subset of an F-normable topological vector space is admissible. C.Krauthausen has given an example of a locally convex subset in KL_ϕ [27] and some sufficient conditions for local convexity of subsets of topological vector spaces.

Let E be a linear space over the real or complex number field. A function $\|\cdot\|^* : E \to [0,\infty)$ will be called a paranorm if and only if:

1. $\|x\|^* = 0 \Longleftrightarrow x = 0$.
2. $\|-x\|^* = \|x\|^*$, for every $x \in E$.

3. $\| x+y \|^* \leq \|x\|^* + \|y\|^*$, for every $x,y \in E$.

4. If $\| x_n - x_o \|^* \to 0$, $\lambda_n \to \lambda_o$ then $\| \lambda_n x_n - \lambda_o x_o \|^* \to 0$.

Then $(E, \| \cdot \|^*)$ is a paranormed space. E is also a topological vector space in which the fundamental system of neighbourhoods of zero is given by the family $\{ U_r \}_{r > 0}$ where $U_r = \{x \mid x \in E , \| x \|^* < r \}$.

In [43] Zima has proved the following fixed point theorem, which is, in fact, a generalization of Schauder's fixed point theorem.

THEOREM 2. Let $(E, \| \cdot \|^*)$ be a paranormed space and $K \subseteq E$ a nonempty convex subset of E such that there exists $C(K) > 0$ so that:

(1) $$\| tx \|^* \leq C(K)\, t \, \|x\|^* \quad , \text{ for every } (t,x) \in [0,1] \times (K-K).$$

If $f:K \to K$ is a compact mapping then $Fix(f) \neq \emptyset$.

PROOF. It is easy to prove that the set K is a locally convex and admissible subset of K and so, using the fixed point theorem of Hahn and Pötter, it follows that $Fix(f) \neq \emptyset$.

DEFINITION 3. Let $(E, \| \cdot \|^*)$ be a paranormed space and $K \subseteq E$ a nonempty subset of E such that (1) holds for some $C(K) > 0$. Then we say that K satisfies the Zima condition.

Using the simplex method, similarly as in Tychonoff's fixed point theorem, Rzepecki has proved in [39] the following fixed point theorem:

THEOREM 3. Let E be a topological vector space, K a nonempty closed and convex subset of E, Z a compact, locally convex subset of K and $f:K \to Z$ a continuous mapping. Then $Fix(f) \neq \emptyset$.

It is easy to see that Zima's fixed point theorem is a Corollary of Rzepecki's fixed point theorem.

Now, we shall give also an example of locally convex, admissible subsets of a class of topological vector spaces, using the notations from Kasahara's paper [22]. A linear mapping Φ of a topological semifield E into another F is said to be positive if $\Phi(x) \geq 0$ in F , for every $x \in E$ with $x \geq 0$. Let $\| \cdot \|$ be a mapping of a linear space X into a topolgical semifield E and let Φ be a continuous positive linear mapping of E into itself. The triplet $(X, \| \cdot \|, \Phi)$ is called a Φ-paranormed space over E and $\| \cdot \|$ a Φ-paranorm on X over E if and only if the following conditions are satisfied:

(P1) $\|x\| \geq 0$, for every $x \in X$.

(P2) $\|tx\| = |t| \cdot \|x\|$, for every $t \in K$ and every $x \in X$

($K \in \{R,C\}$)

(P3) $\|x+y\| \leq \Phi(\|x\| + \|y\|)$, for every $x,y \in X$.

DEFINITION 4. Let $(X, \|\cdot\|, \Phi)$ be a Φ-paranormed space and $K \subset X$. The set K is of Φ-type if and only if for every $n \in N$, every $x = (x_i) \in \prod_1^n (K-K)$ and every $\lambda = (\lambda_i) \in [0,1]^n$ such that $\sum_{i=1}^n \lambda_i = 1$, we have:

$$\| \sum_{i=1}^n \lambda_i x_i \| \leq \sum_{i=1}^n \lambda_i \, \Phi \, (\|x_i\|) .$$

If $K \subseteq X$ is of Φ-type, where $(X, \|\cdot\|, \Phi)$ is a Φ-paranormed space, it is a locally convex admissible subset of X and so we have the following corollary of Hahn and Pötter fixed point theorem [7] :

COROLLARY 1. Let A be a compact and convex subset of Φ-type of a Φ-paranormed space $(X, \|\cdot\|, \Phi)$ and $h: A \to A$ a continuous mapping. Then $Fix(h) \neq \emptyset$.

REMARK. In [22] Kasahara has proved the following result:

Let (X, τ) be a topological vector space. Then there exists a Φ-paranormed space $(X, \|\cdot\|, \Phi)$ over a Tychonoff semifield E such that:

(1) For every neighbourhood U of $0 \in X$ there are an $\varepsilon > 0$ and an indecomposable idempotent $\rho \in E$ such that :

$$\{ x \mid x \in X , \|x\| \cdot \rho \leq \varepsilon \rho \} \subset U .$$

(2) For every neighbourhood U of $0 \in E$ the set:

$$\{ x \mid x \in X , \|x\| \in U \}$$

is a neighbourhood of $0 \in X$.

The following problem is closely related to a fixed point theorem of Lifschitz-Sadovski's type [27]. Let K be a convex, locally convex subset of an F-space. Is the following implication valid:

$$A \subseteq K, \ A \text{ is compact} \Rightarrow \text{co } A \text{ is relatively compact?}$$

An example of such a set $K \subseteq X$, where X is a non locally convex topological vector space, is given in [27].

DEFINITION 5. Let X be a topological vector space, K a nonempty subset of X and $f: K \to X$. The mapping f is generalized condensing (GC) if and only if :

(a) f is conitunous

(b) $A \neq \emptyset$, $A \subset K$, $f(A) \subset A$, $A \setminus \overline{co}\, f(A)$ is compact imply that $\overline{A}$ is compact.

Then, we have the following fixed point theorem [27].

THEOREM 4. Let X be an F-space, K a closed, convex, locally convex subset of X so that :

$$A \subset K,\ A \text{ is compact} \Rightarrow co\, A \text{ is relatively compact.}$$

If f is a generalized condensing mapping from K into K then $Fix(f) \neq \emptyset$.

The proof is similar to that given by V.Stallbohm in [42].

DEFINITION 6. [14]. Let X be a topological vector space, $\mathcal{U}$ a fundamental system of neighbourhoods of zero in X and $K \subseteq X$. The set K is of Z-type if and only if for every $V \in \mathcal{U}$ there exists $U \in \mathcal{U}$ so that $co(U \cap (K-K)) \subseteq V$.

It is easy to see that every set K which satisfies the Zima condition is of Z-type and every set of Φ-type is of Z-type [6]. In a locally convex space every subset is of Z-type since we can suppose that $co(U) = V$. Every convex subset of Z-type is a locally convex subset. It is proved in [10] that for every subset K of Z-type:

$$A \subseteq K,\ A \text{ is relatively compact} \Rightarrow co\, A \text{ is relatively compact}$$

and in [10] is given a fixed point theorem for GC mappings defined on subsets of Z-type.

Now, we shall prove a generalization of a fixed point theorem from [25].

DEFINITION 7. [25] Let X be a topological space and $F: X \to X$. A subset M of X is said to be an attractor for compact sets under F if and only if:

(i) M is nonempty compact set and $F(M) \subseteq M$.

(ii) For any compact subset $C \subset X$ and any open neighbourhood U of M , there exists an integer m such that :

$$F^n(C) \subseteq U \quad , \text{ for all } \quad n \geq m \ .$$

THEOREM 5. Let X be a topological vector space and $\mathcal{U}$ the family of neighbourhoods of zero in X . Further, let G be a nonempty, complete , convex subset of X of Z-type and $F:G \to G$. If :

(i) $\{F^n\}_{n \in N}$ is an equicontinuous family of functions,

(ii) There exists $M \subseteq G$ which is an attractor for compact sets under F ,

then $\mathrm{Fix}(F) \neq \emptyset$.

PROOF. Let $Y = \overline{\mathrm{co}}\, M$. Since the set G is of Z-type and G is convex it follows that Y is of Z-type too. The set M is compact and $M \subseteq G$ so it follows that the set Y is compact. Let:

$$X = \overline{\bigcup_{n=0}^{\infty} F^n(Y)} \ , \ F^0(Y) = Y \ .$$

Then it follows [25] that the set X is compact. Let $A = \bigcap_{n=1}^{\infty} F^n(X)$. Then A is a retract of X [25]. Let $g = F\, r$. g is a continuous mapping from Y into Y . Since G is of Z-type it is a locally convex subset and we can apply Rzepecki's fixed point theorem which implies that $\mathrm{Fix}(g) \neq \emptyset$. Then, as in [25], it follows that $\mathrm{Fix}(F) \neq \emptyset$.

2. FIXED POINT THEOREMS FOR MULTIVALUED MAPPINGS IN NOT NECESSARILY LOCALLY CONVEX TOPOLOGICAL VECTOR SPACE

There are a great number of fixed point theorems for mutivalued mappings in a topological vector space X if X is locally convex. A very complete survey of such theorems can be found in [15] and [16]. Some fixed point theorems for multivalued mappings in not necessarily locally convex topological vector space are proved in [16], [21], [8] and [9]. Here, we shall give some results of the author.

First, we shall give some notations and a definition. Let X and Y be topological vector spaces, $\mathcal{V}$ the family of all neighbourhoods of zero in Y, $\mathcal{U}$ the family of all neighbourhoods of zero in X and 2^Y the family of all non-empty subsets of Y. For $V \in \mathcal{V}$ and $A \subseteq Y$ let:

$$V[A] = \{ y \mid y \in Y,\ z \in A,\ z - y \in V \}.$$

DEFINITION 8. <u>A mutivalued mapping</u> $F: K \to 2^Y$ $(K \subseteq X)$ <u>is uniformly u-continuous if for every</u> $V \in \mathcal{V}$ <u>there exists</u> $W \in \mathcal{U}$ <u>such that</u> $x_1 - x_2 \in W$ $(x_1, x_2 \in K)$ <u>implies</u> $F(x_1) \subseteq V[F(x_2)]$ <u>and</u> $F(x_2) \subseteq V[F(x_1)]$.

Some properties of uniformly u-continuous multivalued mappings are investigated in [31]. In [9] the following theorem on almost continuous selection property is proved. We suppose that every $W \in \mathcal{U}$ is open.

THEOREM 6. <u>Let</u> X <u>be a topological vector space</u>, K <u>a paracompact convex subset of</u> X <u>and</u> $F: K \to R(K)$ <u>a uniformly u-continuous mapping where</u> $R(K)$ <u>is the family of all closed and convex subsets of</u> K. <u>If</u> $F(K)$ <u>is of Z-type then for every</u> $V \in \mathcal{U}$ <u>there exists a continuous mapping</u> $g_V: K \to K$ <u>so that</u>:

$$g_V(x) \in F(x) + V \quad (x \in K).$$

PROOF. Let $V \in \mathcal{U}$. Since the set $F(K)$ is of Z-type there exists $U \in \mathcal{U}$ such that $co(U \cap (F(K) - F(K))) \subseteq V$. Since the mapping F is uniformly u-continuous there exists $W \in \mathcal{U}$ such that:

$$x_1, x_2 \in K,\ x_1 - x_2 \in W,\ y_1 \in F(x_1) \Rightarrow \text{there exists } y_2 \in F(x_2)$$
$$\text{and } y_1 - y_2 \in U.$$

Let $\mathcal{P}$ be the locally finite partition of unity subordinated to the open cover $\{x+W\}_{x\in K}$ and let $h: \mathcal{P}\to K$ be such that $p(x) = 0$ for all $x\in K\setminus\{h(p) + W\}$. Furthermore, let z be a choice function for the family $\{F(x)\}_{x\in K}$ and for every $x\in K$:

$$g_V(x) = \sum_{p\in\mathcal{P}} p(x)\ z(h(p)) .$$

It is obvious that g_V is a continuous mapping from K into K. For every $x\in K$ let $M(x) =\{p|p\in\mathcal{P},\ p(x)\neq 0\}$. For every $x\in K$ the set $M(x)$ is finite. We can suppose that every $W\in\mathcal{U}$ is symmetric and so $p\in M(x)$ implies $h(p)-x\in W$. Let $v_p(x)\in F(x)$ be such that $z(h(p))-v_p(x)\in U$. Since $z(h(p))\in F(h(p))$ and the mapping F is uniformly u-continuous such element $v_p(x)$ exists.

If for every $x\in K$ and every $p\in\mathcal{P}$.

$$u_p(x) = \begin{cases} v_p(x) & p\in M(x) \\ z(x) & p\in\mathcal{P}\setminus M(x) \end{cases}$$

and $s(x) = \sum_{p\in\mathcal{P}} p(x)u_p(x)$, for every $x\in K$ it is easy to see that $s(x)\in F(x)$ and $g_V(x)\in s(x) + V\subset F(x) + V$, for every $x\in K$.

COROLLARY 2. <u>Let</u> X <u>and</u> F <u>be as in Theorem 6 and</u> K <u>a compact subset of</u> X. <u>Then for every</u> $V\in\mathcal{U}$ <u>there exists</u> $x_V\in K$ <u>such that</u> $x_V\in F(x_V)+V$.

PROOF. For every $V\in\mathcal{U}$ let $g_V:K\to K$ be a continuous mapping such that $g_V(x)\in F(x)+V$, for every $x\in K$. Since K is compact $\mathcal{P}$ is finite and $g_V:\overline{co}\ S\to\overline{co}\ S$, where $S = \{z(h(p))|p\in\mathcal{P}\}$. From Brouwer fixed point theorem it follows that there exists $x_V\in K$ such that $x_V = g_V(x_V)$ and so $x_V\in F(x_V)+V$.

Now, we shall give a theorem about the fixed approximation property [9].

THEOREM 7. <u>Let</u> X <u>be a topological vector space</u>, K <u>a nonempty, closed and convex subset of</u> X <u>and</u> $F:K\to R(K)$ <u>a closed mapping such that</u> $\overline{F(K)}$ <u>is compact. If</u> $F(K)$ <u>is of</u> Z-<u>type then for every</u> $V\in\mathcal{U}$ <u>there exists a closed finite dimensional mapping</u> $F_V:K\to R(K)$ <u>such that</u> $F_V(K)$ <u>is relatively compact and for every</u> $x\in K$: $F_V(x)\subseteq F(x) + V$.

PROOF. Let $V\in\mathcal{U}$ and $W\in\mathcal{U}$ such that $\overline{W}\subset V$. Since the set $F(K)$ is of Z-type there exists $U\in\mathcal{U}$ such that $co(U\cap(F(K)-F(k)))\subseteq W$ and so

$$\overline{co}(U\cap(F(K)-F(K)))\subseteq\overline{W}\subset V .$$

Since $F(K)$ is relatively compact there exists a finite set $S = \{x_1,x_2,\dots,x_n\}\subseteq F(K)$ such that $F(K)\subset\bigcup_{i=1}^{n}\{x_i+U\}$. Then the mapping $F_V:K\to R(K)$ is defined in the following way:

$$F_V(x) = [F(x) + \overline{co}(U\cap(F(K)-F(K)))]\cap\overline{co}\, S\ ,\ x\in K .$$

It is easy to see that the mapping F_V is a compact, closed and finite dimensional mapping such that $F_V(x)\subseteq F(x)+V$, for every $x\in K$.

Using the above theorem we obtain the following fixed point theorem.

THEOREM 8. <u>Suppose that</u> X,K <u>and</u> F <u>are as in Theorem 7. Then there exists</u> $x\in K$ <u>such that</u> $x\in F(x)$.

PROOF. It follows from Theorem 7 that for every $V\in\mathcal{U}$ there exists $F_V:\overline{co}\, S\to\overline{co}\, S$, which satisfies all the conditions of Kakutani's fixed point theorem and $F_V(x)\subseteq F(x)+V$, for every $x\in\overline{co}\, S$. So, for every $V\in\mathcal{U}$ there exists $x_V\in\overline{co}\, S$ such that $x_V\in F_V(x_V)$. Using the compactness of the set $\overline{co}\, S$ it follows that there exists a convergent subnet $\{x_W\}$ of the net $\{x_V\}$ such that $\lim_W x_W = x\in\overline{co}\, S$. Then $x\in F(x)$.

Now, we can prove a generalization of Theorem 13 from [1].

THEOREM 9. <u>Let, for every</u> $i\in I$, K_i <u>be a nonempty, compact and convex subset of a topological vector space</u> E_i, $K = \prod_{i\in I} K_i$ <u>and for every</u> $i\in I$, $K_i' = \prod_{j\neq i} K_j$. <u>Further, let for every</u> $i\in I$ S_i <u>be a closed subset of</u> K <u>and suppose that for every</u> $x\in K$ <u>and every</u> $i\in I$ <u>the set</u> $S_i(x)$ <u>is a nonempty and convex subset of</u> K_i, <u>where:</u>

$$S_i(x) =\{ y_i\,|\,y_i\in K_i\ ,[y_i,\hat{x}_i]\in S_i\}\ ,\ \hat{x}_i =\mathrm{proj}_{K_i'}x,\ [y_i,\hat{x}_i] = z\in K$$

<u>and</u> $z_j = y_i$, <u>for</u> $j=i$, $z_j=x_j$, <u>for</u> $j\neq i$ $(x=(x_i))$. <u>If</u> $K_i (i\in I)$ <u>is of</u> Z-<u>type then</u> $\bigcap_{i\in I} S\neq\emptyset$.

PROOF. The proof is similar to the proof of Theorem 13 from [1]. Let us define the mapping $F:K\to R(K)$ in the following way:

$$y\in F(x)\,(x\in K) \iff y = (y_i),\ y_i\in S_i(x)\ ,\ \text{for every}\ i\in I .$$

This means that $F(x) = \prod_{i\in I} S_i(x)$, for every $x \in K$. Let us prove that the mapping F satisfies all the conditions of Theorem 8. Since $S_i(x) = \pi_1(\pi_2^{-1}(\hat{x}_i) \cap S_i)$, for every $x \in K$, where :

$$\pi_1 : K_i \times K_i' \to K_i \quad \text{and} \quad \pi_2 : K_i \times K_i' \to K_i',$$

it follows that $S_i(x)$ is compact for every $x \in K$ and every $i \in I$ and so the set $F(x)$ is closed for every $x \in K$. Since $\operatorname{co} F(x) = F(x)$, for every $x \in K$, we can conclude that $F(x) \in R(K)$, for every $x \in K$. As in [1] it follows that the mapping F is closed. Now, we shall prove that the set K is of Z-type. Let us denote by $\mathcal{V}$ the fundamental system of neighbourhoods of zero in E in the Tychonoff product topology, and by $\mathcal{V}_i$ the fundamental systems of neighbourhoods of zero of E_i, where $E = \prod_{i\in I} E_i$. If $V \in \mathcal{V}$ we shall show that there exists $U \in \mathcal{V}$ so that

$$\text{(2)} \qquad \operatorname{co}(U \cap (K-K)) \subseteq V .$$

Suppose that $V \in \mathcal{V}$. Then there exists a finite set $\{i_1, i_2, \ldots, i_n\} \subseteq I$ such that

$$V = \prod_{i \in I} E_i'$$

where $E_i' = E_i$ if $i \in I \setminus \{i_1, i_2, \ldots, i_n\}$ and $E_i' = V_i \in \mathcal{V}_i$ if $i \in \{i_1, i_2, \ldots, i_n\}$. Since $K_i \subseteq E_i$, for every $i \in I$ and K_i is of Z-type, there exists $U_i \in \mathcal{V}_i$ such that:

$$\operatorname{co}(U_i \cap (K_i - K_i)) \subseteq V_i , \quad i \in \{i_1, i_2, \ldots, i_n\}$$

It is easy to see that the relation (2) is satisfied if the set U is defined by:

$$U = \prod_{i \in I} E_i'', \text{ where } E_i'' = \begin{cases} E_i & i \in I \setminus \{i_1, i_2, \ldots, i_n\} \\ U_i & i \in \{i_1, i_2, \ldots, i_n\} \end{cases} .$$

Now, we can apply Theorem 8 and so there exists $u \in K$ such that $u \in F(u)$. From this it follows that $[u_i, \hat{u}_i] \in S_i$, for every $i \in I$ which implies that $u \in \bigcap_{i \in I} S_i \neq \emptyset$.

REFERENCES

[1] F.Browder, Fixed point theory of multivalued mappings in topological vector space, Math.Ann.197(1968), 283-301.
[2] T.Dobrowolski,H.Tournczyk, On metric linear spaces homeomorphic to ℓ_2 and compact convex sets homeomorphic to Q, Bull. Acad.Po.Sci. 27(1979), 883-887.
[3] M.Ehrlich, Über die Lösbarkeit einer nichtlinearen Integralgeichung in nicht lokalkonvexen Räumen, Math. Nachr.
[4] D.H.Gregory,J.H.Shapiro, Nonconvex linear topologies with the Hahn Banach extension property, Proc. Amer. Math.Soc. 25(1970),902-905.
[5] O.Hadžić, The foundation of the fixed point theory, Dept. of Math. Novi Sad, 1978, 320 p.p. (Serbocroatian).
[6] O.Hadžić, A fixed point theorem in topological vector spaces, Zbornik radova Prirodno-matematičkog fakulteta, Univerzitet u Novom Sadu, serija za matematiku, knjiga 10(1980), 23-29.
[7] O.Hadžić, On the admissibility of topological vector spaces, Acta Sci. Math. 42(1980), 81-85.
[8] O.Hadžić, On multivalued mappings in paranormed spaces, Comm.Math. Univ. Carolinea, 22 ,1(1981), 129-136.
[9] O.Hadžić, Some fixed point and almost fixed point theorems for multivalued mappings in topological vector spaces, Nonlinear Analysis, Theory, Methods & Applications , Vol.5 No.9,(1981), 1009-1019.
[10] O.Hadžić, On Sadovski's fixed point theorem in topological vector spaces, Comm.Math. in print.
[11] O.Hadžić, On Kakutani's fixed point theorem in topological vector spaces, Bull.Acad.Polon.Sci.Sér.Sci.Math. in print.
[12] O.Hadžić, A fixed point theorem for the sum of two mappings, Proc. Amer. Math.Soc. , in print.
[13] O.Hadžić, LJ.Gajić, A fixed point theorem for multivalued mappings in topological vector space, Fund.Math. CIX (1980), 163-167.
[14] O.Hadžić, LJ.Gajić, Some applications of fixed point theorems for multivalued mappings in topological vector space (to appear).
[15] S.Hahn, Fixpunktsätze für mengenwertige Abbildungen in lokalkonvexen Räumen, Math. Nachr. 73 (1976), 269-283.
[16] S.Hahn, A remark on a fixed point theorem for condensing set-valued mappings, Technische Universität Dresden, Informationen, Sektion Mathematik, 07-5-77.
[17] S.Hahn, Ein elementarer Zugang zur Leray-Schauder-Theorie, Technische Universität Dresden, Informationen, Sektion Mathematik, 07-10-77.
[18] S.Hahn, F.K.Potter, Über Fixpunkte kompakter Abbildungen in topologischen Vektor-Räumen, Stud.Math. 50 (1974),1-16.
[19] S.Hahn, T.Riedrich, Der Abbildungsgrad kompakter vektorfelder in nicht notwendig lokalkonvexen topologischen Räumen, Wiss.Z.Techn. Univ. Dresden, 22 (1973), 37-42.
[20] J.Ishii, On the admissibility of function spaces, J.Fac.Sci. Hokkaido Univ. Ser.I, 19 (1965), 49-55.
[21] W.Kaballo, Zum Abbildungsgrad in Hausdorffschen topologischen Vektorräumen, Manuscripta math. 8(1973), 209-216.
[22] S.Kasahara, On formulations of topological linear spaces by topological semifield, Math. Japonicae 19(1974),121-134.

[23] V.Klee, Shrinkable neighbourhoods in Hausdorff linear spaces, Math. Ann. 141(1960), 281-285.
[24] V.Klee, Leray-Schauder theory without local convexity, Math.Ann. 141(1960), 286-296.
[25] H.M.Ko,K.K.Tan, Attractors and a fixed point theorem in locally convex spaces, Comm.Math.Univ. Carolinea, 21,1(1980), 71-79.
[26] C.Krauthausen, On the theorems of Dugundji and Schauder for certain nonconvex spaces, Math. Balk. 4(1974), 365-369.
[27] C.Krauthausen, Der Fixpunktsatz von Schauder in nicht notwendig konvexen Räumen sowie Anwendungen auf Hammersteinsche Gleichungen, Dissertation, 1976,Aachen.
[28] M.Landsberg, Lineare topologische Räume die nicht lokal konvex sind, Math. Zeitschr. 65(1956), 113-132.
[29] M.Landsberg, Über die Fixpunkte kompakten Abbildungen, Math.Ann, 154 (1964), 427-431.
[30] M.Landsberg, T.Riedrich, Über positive Eigenwerte kompakter Abbildungen in topologischen Vektorräumen, Math.Ann, 163(1966), 50-61.
[31] A.Lechicki, On continuous and measurable multifunctions, Comm. Math. Prace Matematyczne, 21(1980), 141-156.
[32] V.I.Lomonosov, Über invariante Teilräume der Menge der Operatoren die mit einem vollstetigen Operator kommutieren, Funkcionalnij analiz i jego priloshenija, t.7 wyp. 3, (1973), 55-56 (Russian).
[33] V.E.Matusov, Obobščenie teoremy o nepodvižnoi točke Tihonova, Doklady A.N.Uz SSR,No.2(1978), 12-14(Russian).
[34] E.Michael, Continuous Selection I, Ann. of Math. 63 (1956),361--382.
[35] J.Reinermann, V.Stallbohm, Fixed point theorems for compact and nonexpansive mappings on starshaped domains, Mathematica Balkanica, 4(1974), 511-516.
[36] T.Riedrich, Die Räume $L^p(0,1)$ $(0 < p < 1)$ sind zulässig, Wiss.Z. Techn. Univ. Dresden, 12(1963), 1149-1152.
[37] T.Riedrich, Der Räum S(0,1) ist zulässig, Wiss.Z.Techn. Univ. Dresden, 13(1964), 1-6.
[38] T.Riedrich, Das Birkhoff Kellogg-theorem für lokal radial beschränkte Räume, Math. Ann. 166(1966), 264-276.
[39] B.Rzepecki, Remarks on Schauder's fixed point theorem, Bull. Acad. Polon. Sci.Sér. Sci.Math.Astronom.Phy., 24(1976), 589-603.
[40] T.A.Sarimsakov, Novoe dokazatel'stvo teoremy Tihonova, U.M.N. 20 (124),(1965).
[41] E.Schulz, Existenzreihe für Halbeigenwerte kompakten Abbildungen in topologischen Vektorräumen, Math.Nachr. 57 (1973), 182-199.
[42] V.Stallbohm, Fixpunkte nichtexpansiver Abbildungen, Fixpunkte kondensierender Abbildungen, Fredholm'sche Sätze linearer kondensierender Abbildungen, Dissertation an der R.W.T. H. Aachen, 1973.
[43] K.Zima, On the Schauder's fixed point theorem with respect to paranormed space, Comm.Math. 19(1977),421-423.

INTRODUCTION TO LIE GROUPS AND THEIR REPRESENTATIONS

Alain Guichardet

I. INTRODUCTION: QUANTUM MECHANICAL OPERATORS, GALILEAN LIE ALGEBRA AND GALILEAN LIE GROUP

Consider the physical system consisting of one single particle moving freely (i.e. without forces) in the space R^3 ; the Quantum Mechanics associates to this system a complex Hilbert space $\mathcal{H} = L^2(R^3)$, and a number of operators representing several physical observables; following the principles of Quantum Mechanics due to von Neumann, these operators have to be self-adjoint and in particular one must define their domain with a great care; however in this Introduction we will define these operators only formally.

We have first 3 operators Q_1, Q_2, Q_3 representing the (3 components of the) position of our particle ; Q_j is the multiplication operator by the function $R^3 \ni x = (x_1, x_2, x_3) \to x_j$, i.e.

$$(Q_j \cdot \psi)(x) = x_j \cdot \psi(x) \quad , \quad \psi \in \mathcal{H} .$$

We then have 3 operators P_1, P_2, P_3 representing the impulse of our particle and defined by $P_j = -i\hbar \frac{\partial}{\partial x_j}$, i.e.

$$(P_j \cdot \psi)(x) = -i\hbar \frac{\partial \psi}{\partial x_j}$$

where $\hbar$ is equal to Planck constant divided by 2π .

We then encounter 3 operators L_1, L_2, L_3 representing the orbital kinetic momentum, defined by

$$L_1 = Q_2 \cdot P_3 - Q_3 \cdot P_2$$

and so on by circular permutations.

Finally we have the hamiltonian operator H , representing the energy of the particle (also called "free hamiltonian" since there is no potential)

$$H = -\frac{\hbar^2}{2m} \Delta = \frac{1}{2m} \sum_{j=1}^{3} P_j^2 \quad ;$$

here Δ is the Laplace operator.

These 10 operators satisfy the following commutation relations (we write $[A,B]$ for $AB - BA$)

$$[Q_j,Q_k] = [P_j,P_k] = 0 \quad ,$$

$$[Q_j,P_k] = i\hbar\,\delta_{jk} \quad \text{(where } \delta_{jk} = 1 \text{ if } j=k \text{ and } 0 \text{ otherwise)},$$

$$[L_1,L_2] = i\hbar L_3 \quad \text{and circular permutations},$$

(I.1)

$$[L_1,Q_1] = 0\ , [L_1,Q_2] = i\hbar Q_3\ , [L_1,Q_3] = i\hbar Q_2 \quad \text{and circular permutations},$$

$$[L_1,P_1] = 0\ , [L_1,P_2] = i\hbar P_3\ , [L_1,P_3] = -i\hbar P_2 \quad \text{and circular permutations},$$

$$[Q_j,H] = \frac{i\hbar}{m} P_j \quad ,$$

$$[P_j,H] = [L_j,H] = 0 \ .$$

The commutation relations (I.1) ("Heisenberg commutation relations") are of special importance since they lead to the "Heisenberg uncertainty relations" linking position and impulse. We have written the above commutation relations only formally, without worrying about questions of definition domains for the operators; one can also consider these relations from a purely abstract point of view, as linking 10 objects defined purely abstractly; but it is more efficient to consider these 10 objects as generating a vector space (more precisely as forming a basis of a vector space) and to extend the commutation relations by linearity to this vector space; we shall obtain a new mathematical object called a Lie algebra : a finite dimensional real vector space E with a bilinear internal operation $(X,Y) \to [X,Y]$ satisfying (instead of associativity relation) the Jacobi identity

$$[X,[Y,Z]]+[Y,[Z,X]]+[Z,[X,Y]] = 0 \ .$$

In the case at hand it is more or less natural to introduce the Galilean Lie algebra (i.e.the Lie algebra of the Galilean group which is the invariance group of the classical mechanics). This Lie algebra g is the set of all 5×5 - real matrices which have the form

$$X = \begin{pmatrix} \alpha & \beta & \gamma \\ 0 & 0 & \delta \\ 0 & 0 & 0 \end{pmatrix} ,$$

where

a) $\alpha \in so(3)$ = the set of antisymmetric 3×3 -matrices ,

b) β , γ are column matrices with 3 entries,

c) δ is a real number.

We shall write $X = (\alpha, \beta, \gamma, \delta)$.

We define $[X,X'] = XX' - X'X$ and we obtain

$$[X,X'] = ([\alpha, \alpha'], \alpha\beta' - \alpha'\beta, \alpha\gamma' - \alpha'\gamma + \beta\delta' - \beta'\delta, 0).$$

We define the following basis of g : for ℓ_1, ℓ_2, ℓ_3 the β, γ and δ-parts are zero, and the α-parts are respectively

$$\begin{pmatrix} 0 & 0 & 0 \\ 0 & 0 & -1 \\ 0 & 1 & 0 \end{pmatrix} , \begin{pmatrix} 0 & 0 & 1 \\ 0 & 0 & 0 \\ -1 & 0 & 0 \end{pmatrix} , \begin{pmatrix} 0 & -1 & 0 \\ 1 & 0 & 0 \\ 0 & 0 & 0 \end{pmatrix} ;$$

for q_1, q_2, q_3 the α, γ and δ-parts are zero and the β-parts are respectively

$$\begin{pmatrix} 1 \\ 0 \\ 0 \end{pmatrix} , \begin{pmatrix} 0 \\ 1 \\ 0 \end{pmatrix} , \begin{pmatrix} 0 \\ 0 \\ 1 \end{pmatrix} ;$$

for p_1, p_2, p_3 it is the same with β being replaced by γ ; finally for h the α, β and γ-parts are zero and $\delta = 1$.

The commutation relations between these 10 basis elements are the following

$$[q_j, q_k] = [p_j, p_k] = 0 ,$$

$$[q_j, p_k] = 0 ,$$

$[\ell_1, \ell_2] = \ell_3$ and circular permutations,

$[\ell_1, q_1] = 0$, $[\ell_1, q_2] = q_3$, $[\ell_1, q_3] = -q_2$ and circular permutations,

$$[\ell_1, p_1] = 0 \ , [\ell_1, p_2] = p_3 \ , [\ell_1, p_3] = -p_2$$ and circular permutations,

$$[q_j, h] = p_j \ ,$$

$$[p_j, h] = [\ell_j, h] = 0 \ .$$

We note a strong similarity with the relations between the quantum mechanical operators, with one very important difference:

$$[q_j, p_k] = 0 \qquad , [Q_j, P_k] = i \hbar \delta_{jk} \ .$$

Let us try to construct a representation of g in H , i.e. a linear mapping U of g into the operators in H , conserving the brackets ; the best we can do is to put

$$U(q_j) = - i m Q_j$$

$$U(p_j) = \frac{i}{\hbar} P_j$$

$$U(\ell_j) = - \frac{i}{\hbar} L_j$$

$$U(h) = \frac{i}{\hbar^2} H \ .$$

However, $[U(X), U(X')] - U([X,X'])$ is not equal to zero, but to the scalar operator $i \sigma(X,X')$, where

$$\sigma(X,X') = m(\beta \cdot \gamma' - \beta' \cdot \gamma) \ ,$$

where, for $x,y \in R^3$, $x \cdot y$ denotes the usual scalar product $x_1 y_1 + x_2 y_2 + x_3 y_3$.

If one insists to get $[U,U] - U([\cdot,\cdot])$ equal to 0 , one has to replace g by another Lie algebra

$$\tilde{g} = g \oplus R$$

with the following bracket operation :

$$[(X,\xi),(X',\xi')] = ([X,X'] , \sigma(X,X')), \qquad X \in g , \xi \in R$$

and to extend U as follows

$$\tilde{U}(X,\xi) = U(X) + i \xi I \ ;$$

now we really get a representation $\tilde{U}$, but of $\tilde{g}$ and not of g which appears as a quotient of $\tilde{g}$, and not as a subalgebra.

REMARK I.1. This situation is a particular case of the "extension theory" of Lie algebras : σ is a 2-cocycle, which means that

$$\sigma([X,Y],Z) + \sigma([Y,Z],X) + \sigma([Z,X],Y) = 0 \ ;$$

thanks to this relation, $\tilde{g}$ is a Lie algebra which contains R as an ideal (here this ideal is central), the quotient being g ; one says that $\tilde{g}$ is an <u>extension</u> of g by R, and that U is a <u>projective represenation</u> of g with <u>multiplier</u> σ.

REMARK I.2. One can say that the commutation relations between the quantum mechanical operators are "abstractly contained" in the Lie algebra $\tilde{g}$; actually there are other relations between these operators; for instance if one sets $L^2 = L_1^2 + L_2^2 + L_3^2$ one has

$$L^2 \cdot L_j = L_j \cdot L^2 \ , \qquad j = 1,2,3 \ ;$$

such relations, containing ordinary products, are not "abstractly contained" in $\tilde{g}$ (a Lie algebra contains only bracket-products), but in an associative algebra associated to $\tilde{g}$ and called its "enveloping algebra".

REMARK I.3. It will be more appropriate to replace the operators Q_j, P_j, L_j, H by their Fourier transforms (i.e. to pass to the momentum-space); precisely define the Fourier transform of a function ψ by

$$\hat{\psi}(p) = (2\pi)^{-\frac{3}{2}} \int e^{-ip\cdot x} \psi(x)\, dx \ ;$$

then our operators become

$$\hat{Q}_j = i\frac{\partial}{\partial p_j} \ , \qquad \hat{P}_j = \hbar\, p_j$$

$$\hat{H} = \frac{\hbar^2}{2m}|p|^2 \qquad (\text{where } |p|^2 = p\cdot p = p_1^2 + p_2^2 + p_3^2)$$

$$\hat{L}_1 = \hat{Q}_2 \cdot \hat{P}_3 - \hat{Q}_3 \cdot \hat{P}_2 \ , \ \text{etc.}$$

Clearly, we have the same commutation relations between the $\hat{Q}_j, \ldots$, as between the $Q_j, \ldots$.

The purpose of the following lectures is to give firstly some general information about Lie groups and Lie algebras, representation theory and in particular Mackey theory for semi-direct products, and then to apply them to obtain a classification of representations of the Galilean group which is the set of all 5×5 matrices of the form

$$g = \begin{pmatrix} A & B & C \\ 0 & 1 & D \\ 0 & 0 & 1 \end{pmatrix},$$

where

a) $A \in SO(3)$ = the set of orthogonal 3×3 matrices with determinant 1 ,

b) B,C are column matrices with 3 entries,

c) D is a real number .

II. LIE GROUPS AND THEIR LIE ALGEBRAS

II.1. General definitions. We shall give a definition which is more resrictive than the usual one, but which is large enough for many purposes.

Denote by k a field equal to R or C, and by V a vector space over k which has a finite dimension n ; one denotes by End V or $M(n,k)$ the set of k-linear mappings in V or the set of all $n \times n$ matrices with entries in k, by $GL(V)$ or $GL(n,k)$ the subset consisting of invertible mappings or matrices. The vector space End V is a Lie algebra for the bracket

$$[X,Y] = XY - YX \quad .$$

We recall the exponential mapping exp defined by

$$\exp X = \sum_{m=0}^{\infty} \frac{X^m}{m!}$$

(the series is absolutely convergent for any norm) ; it sends End V into $GL(V)$, it is C^∞ (i.e. infinitely differentiable) and its differential at 0 is I ; it induces a diffeomorphism of a neighborhood U_o of 0 in End V onto a neighborhood V_o of I in $GL(V)$, with inverse mapping log given by

$$\log (I+T) = \sum_{m=1}^{\infty} (-1)^{m+1} \frac{T^m}{m} \quad .$$

One has $\exp(X + Y) = \exp X \cdot \exp Y$ if X and Y commute ; in particular

$$\exp(s+t) X = \exp sX \cdot \exp tX \quad , \qquad s,t \in R \ .$$

DEFINITION II.1. A Lie group is a closed subgroup G of some $GL(n,k)$, $n \in N^*$, $k = R$ or C.

It is a theorem of von Neumann that such a group G is a submanifold of $M(n,k)$ (which we shall always consider as a real vector space) ; this means that there exists a neighborhood V of I in $GL(n,k)$ and a finite number of real functions $F_1,\ldots,F_p$ on V, which are C^∞ with respect to the coefficients a_{ij} of the variable $A \in V$ (resp. with respect to the real and imaginary parts a'_{ij}, a''_{ij} of the a_{ij}'s) if $k = R$ (resp. if $k = C$), such that, for $A \in V$, we have

$$A \in G \Leftrightarrow F_1(A) = F_2(A) = \dots = F_p(A) = 0 .$$

We denote by $\mathcal{g}$ the tangent subspace to G at I (more precisely the vector subspace parallel to it) ; this is the set of all $X \in M(n,k)$ satisfying

$$\sum_{i,j} \frac{\partial F_q}{\partial a_{ij}}(I) \cdot x_{ij} = 0 \qquad \forall q \quad \text{if} \quad k = R ,$$

$$\sum_{i,j} \left(\frac{\partial F_q}{\partial a'_{ij}} (I) \cdot x'_{ij} + \frac{\partial F_q}{\partial a''_{ij}} (I) \cdot x''_{ij} \right) = 0 \quad \forall q \quad \text{if} \quad k = C .$$

We shall consider $\mathcal{g}$ as a <u>real</u> vector space (vector subspace of the real vector space $M(n,k)$) .

II.2. EXAMPLES.

a) $G = GL(n,k)$, $\mathcal{g} = M(n,k)$, also denoted by $\mathcal{g}\ell(n,k)$.

b) $G = SL(n,k) = \{g \in GL(n,k) \mid \det g = 1 \}$,

$\mathcal{g} = s\ell(n,k) = \{ X \in \mathcal{g}\ell(n,k) \mid \operatorname{Tr} X = 0 \}$.

c) $G = O(n,k) = \{g \in GL(n,k) \mid g \cdot {}^t g = I\}$ = orthogonal group,

$\mathcal{g} = o(n,k) = \{X \in \mathcal{g}\ell(n,k) \mid X + {}^t X = 0 \}$
= the set of antisymmetric matrices.

d) $G = SO(n,k) = SL(n,k) \cap O(n,k)$,

$\mathcal{g} = so(n,k) = o(n,k)$.

One writes $SO(n), \dots,$ instead of $SO(n,R), \dots$.

e) $G = U(n) = \{g \in GL(n,c) \mid g \cdot {}^t\bar{g} = I\}$ = unitary group ,

$\mathcal{g} = u(n) = \{X \in \mathcal{g}\ell(n,C) \mid X + {}^t\bar{X} = 0 \}$

= the set of complex antihermitian matrices.

f) $G = SU(n) = SL(n,C) \cap U(n)$,

$\mathcal{g} = su(n) = s\ell(n,c) \cap u(n)$.

g) R^n is isomorphic to the group of matrices $g \in GL(n+1,R)$ of the form

$$g = \begin{pmatrix} 1 & 0 & \cdots\cdots & 0 & x_1 \\ 0 & 1 & \cdots\cdots & 0 & x_2 \\ \cdot & \cdot & & \cdot & \cdot \\ \cdot & \cdot & & \cdot & \cdot \\ \cdot & \cdot & & \cdot & \cdot \\ 0 & 0 & \cdots\cdots & 1 & x_n \\ 0 & 0 & \cdots\cdots & 0 & 1 \end{pmatrix} \quad ;$$

then $\mathcal{g}$ is similar with 0's replacing 1's .

h) The euclidean group (or rigid motion group) in R^n is the set of $(n+1) \times (n+1)$- matrices of the form

$$g = \begin{pmatrix} A & B \\ 0 & 1 \end{pmatrix}$$

with $A \in SO(n)$, $B \in R^n$ (column vector). Then $\mathcal{g}$ is the set of matrices of the form

$$X = \begin{pmatrix} \alpha & \beta \\ 0 & 0 \end{pmatrix}$$

with $\alpha \in \mathfrak{so}(n)$, $\beta \in R^n$.

i) The Galilean group has been described in the Introduction.

II.3. <u>The Lie algebra of a Lie group</u>. Let G be a Lie group as defined in II.1 ; an element X of $\mathcal{g}$ is , by definition, the derivative at 0 of a C^1 mapping γ sending an interval $]-a,a[$ into G and the point 0 into I :

$$\gamma(t) \in G \quad \forall t \ , \ \gamma(0) = I \ , \quad \gamma'(0) = X \ .$$

THEOREM II.1. The subspace $\mathcal{g}$ of $M(n,k)$ is a Lie subalgebra, (i.e. is closed for the $[\cdot,\cdot]$ operation).

PROOF. First consider $X = \gamma'(0)$ as before; for every $g \in G$, the mapping $t \to g \cdot \gamma(t) \cdot g^{-1}$ sends $]-a,a[$ into G and its derivative at 0 is $g \cdot X \cdot g^{-1}$; hence

$$g \cdot X \cdot g^{-1} \in \mathcal{g} \ .$$

Consider now an element $Y \in \mathcal{g}$, derivative of a mapping ζ , we know that $\zeta(t) \cdot X \cdot \zeta(t)^{-1} \in \mathcal{g} \ \forall t$; hence its derivative at 0 also belongs to g ; but this derivative is equal to $[Y,X]$.

REMARK II.1. There are other ways of viewing at g . First of all consider the set $\overline{g}$ of all vector fields on G which are left invariant, i.e. invariant under left translation by elements of the group G ; evaluating such a vector field at the point I we get a linear isomorphism between $\overline{g}$ and g ;moreover if $\overline{X},\overline{Y} \in \overline{g}$ correspond to $X,Y \in g$, the element of $\overline{g}$ which corresponds to $[X,Y]$ is $\overline{X}\,\overline{Y} - \overline{Y}\,\overline{X}$ (here we consider $\overline{X},\overline{Y}$ as differential operators on G).

Secondly, consider the set $\mathcal{U}(g)$ of all distributions (= continuous linear functionals on $C_c^\infty(G)$) supported by the point I ; they are precisely the linear combinations of partial derivatives of the Dirac measure δ ; endowing $\mathcal{U}(g)$ with the operation of convolution, we get an associative algebra which contains g as the set of distributions containing only derivatives of order 1 ; $\mathcal{U}(g)$ is called the <u>enveloping algebra</u> of g .It can also be defined in an algebraic manner, as a quotient of the tensor algebra of g .

II.4. <u>The exponential mapping</u>.

THEOREM II.2. (i) The mapping exp (see definition in II.1) sends g into G and induces a diffeomorphism of a neighborhood U of 0 in g onto a neighborhood V of I in G .

(ii) For every $X \in g$, the mapping $R \ni t \to \exp tX$ is a <u>one-parameter subgroup</u>, i.e. is C^∞ and satisfies

$$\exp (s+t) X = \exp sX \cdot \exp tX \ .$$

(iii) The mapping $t \to \exp tX$ is the only one-parameter subgroup having X as its derivative at 0 .

PROOF. (i) Write $X = \gamma'(0)$ with $\gamma(t) \in G$, $\gamma(0) = I$ as in II.3 ; for a large integer n , $\gamma(\frac{1}{n})$ belongs to V_o (defined in II.1); hence we can write

$$\gamma(\tfrac{1}{n}) = \exp(\log \gamma(\tfrac{1}{n}))$$

$$(\gamma(\tfrac{1}{n}))^n = \exp(n \cdot \log \gamma(\tfrac{1}{n})) \quad ;$$

an elementary limit calculation shows that

$$\lim_{n\to\infty} (n \cdot \log \gamma (\tfrac{1}{n})) = X \quad ;$$

hence

$$\exp X = \lim_{n\to\infty} (\gamma (\tfrac{1}{n}))^n \quad ;$$

the second member belongs to G since $(\gamma(\frac{1}{n}))^n \in G$ and G is closed in GL(n,k) ; this proves that $\exp g \subset G$ .Then (i) results from the facts that exp is a local diffeomorphism and that g and G are manifolds of the same dimension.

(ii) is immediate ,

(iii) Let γ be such a one parameter subgroup ; derivating the relation $\gamma(s+t) = \gamma(s) \cdot \gamma(t)$, we get the differential relation

$$\gamma'(t) = X \cdot \gamma(t) \quad ;$$

we also have the initial condition

$$\gamma(0) = I \ .$$

It then suffices to invoque the unicity theorem for linear differential equations.

The exp mapping is generally speaking neither injective nor surjective, but :

PROPOSITION II.1. If G is connected, every element of G can be written as a product $\exp X_1 \cdots \exp X_m$ with $m \in N$ and $X_1,\ldots,X_m \in g$.

PROOF. The set G' of the elements of that form is a subgroup of G , which is open since it contains a neighborhood of I ; hence every residue class modulo G' is also open, due to the bicontinuity of the translation mapping; hence the union of all classes distinct from G' is also open ; this implies that G' is also closed, hence equal to G since G is connected.

II.5. Morphisms of Lie groups

DEFINITION II.2. Given two Lie groups G and H , a morphism of G into H is a mapping which is both multiplicative and continuous.

One can prove (cf. [1], part II, ch V) that such a morphism is automatically of class C^∞ ; hence we can consider the differential of our morphism ϕ at the point I ; it is a linear mapping $D\phi$ of g into h which is characterized by

$$(D\phi)(X) = \frac{d}{dt} \phi(\exp tX)\Big|_{t=0}, \qquad \forall X \in g .$$

THEOREM II.3. (i) $D\phi$ is a morphism of Lie algebras, i.e. is R-linear and preserves brackets.

(ii) $\phi(\exp X) = \exp(D\phi(X)) \qquad \forall X \in g$

(iii) If G is connected, the mapping $\phi \to D\phi$ is injective.

PROOF. (i) Proof similar to that of theorem II.1 : for $X \in g$, $g \in G$ we have

$$\phi(\exp t\, gXg^{-1}) = \phi(g \cdot \exp tX \cdot g^{-1}) = \phi(g) \cdot \phi(\exp tX) \cdot \phi(g)^{-1}$$

whence, by derivation:

$$D\phi(gXg^{-1}) = \phi(g) \cdot D\phi(X) \cdot \phi(g)^{-1} \quad ;$$

now taking $g = \exp tY$ and derivating, one gets the result.

(ii) follows from the fact that $t \to \phi(\exp tX)$ and $t \to \exp(t \cdot D\phi(X))$ are two one-parameter subgroups of H having $D\phi(X)$ as their derivatives at 0 .

(iii) If G is connected and $D\phi_1 = D\phi_2$, part (ii) shows that ϕ_1 and ϕ_2 coincide on the elements $\exp X$, then on their products, i.e. everywhere by Proposition II.1.

REMARK II.2. One can prove that if G is connected and simply connected, the mapping $\phi \to D\phi$ is bijective: every morphism $g \to h$ can be lifted to (i.e. is the differential of) a morphism $G \to H$.(cf. [2]).

EXAMPLE. When $H = GL(V)$ for some finite dimensional vector space V (and then $h = g\ell(V)$), a morphism of G into H is also called a <u>representation</u> of G <u>in</u> V (for this notion, see §III) and its differential - the <u>differential</u> or <u>infinitesimal representation</u> associated to it.

In particular consider, for $g \in G$, the linear mapping in g $\mathrm{Ad}\, g : X \to g \cdot X \cdot g^{-1}$ (cf. proof of theorem II.1); Ad is a representation of G in g called the <u>adjoint representation</u> ; we shall use it presently to establish a link between the Lie groups SU(2) and SO(3).

<u>Comparison of</u> SU(2) <u>and</u> SO(3). The Lie algebra $\mathfrak{su}(2)$ has a basis consisting of the matrices

$$X_1 = \frac{1}{2}\begin{pmatrix} 0 & -i \\ -i & 0 \end{pmatrix}, X_2 = \frac{1}{2}\begin{pmatrix} 0 & -1 \\ 1 & 0 \end{pmatrix}, X_3 = \frac{1}{2}\begin{pmatrix} -i & 0 \\ 0 & i \end{pmatrix}$$

(the matrices $2\,i\,X_j$ are known in Quantum Mechanics as the Pauli matrices); the brackets are given by

$$[X_1, X_2] = X_3 \qquad \text{and circular permutations ;}$$

the corresponding one-parameter subgroups are the following :

$$\exp t\,X_1 = \begin{pmatrix} \cos\frac{t}{2} & -i\sin\frac{t}{2} \\ -i\sin\frac{t}{2} & \cos\frac{t}{2} \end{pmatrix}$$

$$\exp t\,X_2 = \begin{pmatrix} \cos\frac{t}{2} & -\sin\frac{t}{2} \\ \sin\frac{t}{2} & \cos\frac{t}{2} \end{pmatrix}$$

$$\exp t\,X_3 = \begin{pmatrix} e^{-it/2} & 0 \\ 0 & e^{it/2} \end{pmatrix}.$$

On the other hand the Lie algebra $\mathfrak{so}(3)$ has a basis consisting of the matrices

$$Y_1 = \begin{pmatrix} 0 & 0 & 0 \\ 0 & 0 & -1 \\ 0 & 1 & 0 \end{pmatrix}, Y_2 = \begin{pmatrix} 0 & 0 & 1 \\ 0 & 0 & 0 \\ -1 & 0 & 0 \end{pmatrix}, Y_3 = \begin{pmatrix} 0 & -1 & 0 \\ 1 & 0 & 0 \\ 0 & 0 & 0 \end{pmatrix}$$

with the brackets

$$[Y_1, Y_2] = Y_3 \quad \text{and circular permutations ;}$$

the corresponding subgroups are easy to describe.

One sees that the Lie algebras $\mathfrak{su}(2)$ and $\mathfrak{so}(3)$ are isomorphic ; one can then expect some relation between the Lie groups. We first remark that for every

$$X = x_1 \cdot X_1 + x_2 \cdot X_2 + x_3 \cdot X_3 \in \mathfrak{su}(2)$$

one has

$$\det X = \frac{1}{4} (x_1^2 + x_2^2 + x_3^2) \quad ;$$

hence, for $g \in SU(2)$ the operator $\mathrm{Ad}\, g$ in $\mathfrak{su}(2)$ preserves the quadratic form $x_1^2 + x_2^2 + x_3^2$; in other words the morphism $\mathrm{Ad}: SU(2) \to GL(\mathfrak{su}(2)) = GL(R^3)$ is actually a morphism $SU(2) \to O(3)$. Since $SU(2)$ is connected (see Proposition II.2. below), $\det \mathrm{Ad}\, g$, which can take only the values ± 1 , can actually take only the value 1 , i.e. $\mathrm{Ad}: SU(2) \to SO(3)$. Now a direct computation shows that

$$\mathrm{Ad}(\exp t X_j) = \exp t Y_j \quad , \qquad j = 1,2,3 \quad ;$$

since $SO(3)$ is connected (Proposition II.2. below), Proposition II.1. shows that Ad is surjective. Finally its kernel consists of all matrices in $SU(2)$ commuting with all matrices in $\mathfrak{su}(2)$, and it is easy to see that the only matrices having this property are $\pm I$.

We have thus proved :

THEOREM II.4. There exists a morphism of $SU(2)$ into $SO(3)$, which is surjective, which transforms $\exp t X_j$ into $\exp t Y_j$ for $j = 1,2,3$, and whose kernel is the set $\{\pm I\}$.

PROPOSITION II.2. The groups $SU(2)$ and $SO(3)$ are connected.

PROOF. For $SU(2)$ this follows from the fact that this group is the set of all matrices of the form $\begin{pmatrix} a & b \\ -\bar{b} & \bar{a} \end{pmatrix}$ where $a, b \in C$ and $|a|^2 + |b|^2 = 1$, in other words $SU(2)$ is homeomorphic to the sphere S^3 , which is connected. For $SO(3)$ it is known (theory of the Euler angles) that every rotation $R \in SO(3)$ can be written in the form

$$R = \exp t_1 Y_3 \cdot \exp t_2 Y_1 \cdot \exp t_3 Y_3$$

which proves that R can be linked to I by a continuous curve.

REMARK II.3. Since S^3 is simply connected, the same is true for $SU(2)$, hence $SU(2)$ is a universal covering of order 2 for $SO(3)$. This fact and Remark II.2. explain why the representation theory is simpler for $SU(2)$ than for $SO(3)$ (see III.3. below).

III. THEORY OF GROUP REPRESENTATIONS

III.1. General (=not necessarily unitary) representations. In this section we give general definitions and properties which are mainly important in the case of finite-dimensional representations.

DEFINITION III.1. A representation of a group G in a complex vector space E is a morphism of G into the group $GL(E)$ of bijective linear operators in E. The operator associated to an element g of G will be written $\Pi(g)$ or U_g. (One sometimes considers representations in real vector spaces, like the adjoint representation of a Lie group (see II.5)). A representation is called trivial if $\Pi(g) = I \ \forall g$.

DEFINITION III.2. Two representations $(E_1,\Pi_1),(E_2,\Pi_2)$ are said to be equivalent if there is a linear isomorphism u of E_1 onto E_2 satisfying

$$u \cdot \Pi_1(g) = \Pi_2(g) \cdot u \qquad \forall g \in G \quad ;$$

such a u is called an interwining operator.

We often speak of a "representation" instead of an "equivalence class of representations".

The principal operations on representations are the following

a) If (E,Π) is a representation and F an invariant vector subspace of E (i.e. $\Pi(g)\, x \in F \ \forall x \in F$, $g \in G$), then the mapping $g \to \Pi(g)|_F$ is called a subrepresentation of Π.

b) The contragredient of a representation (E, Π) is the representation ρ in the dual E^* of E defined by

$$\rho(g) = {}^t\Pi(g)^{-1} .$$

c) Let $(E_i, \Pi_i)_{i \in I}$ be an arbitrary family of representations of G; the direct sum is the representation Π in the direct sum $E = \bigoplus E_i$ defined by

$$(\Pi(g)\cdot x)_i = \Pi_i(g)\cdot x_i \ , \quad x = (x_i) \in E \quad ;$$

one writes $\Pi = \bigoplus \Pi_i$.

d) Let (E_1, Π_1) and (E_2, Π_2) be two representations of G; their tensor product is the representation $\Pi = \Pi_1 \otimes \Pi_2$ in the space $E_1 \otimes E_2$ characterized by

$$\Pi(g)(x_1 \otimes x_2) = \Pi_1(g)\, x_1 \otimes \Pi_2(g)\, x_2 \,, \quad x_i \in E_i \,.$$

DEFINITION III.3. A representation is called <u>irreducible</u> if E contains no other invariant subspaces than $\{0\}$ and E.

REMARK III.1. A representation (even finite dimensional) is not necessarily a direct sum of irreducible subrepresentations ; look for instance at the representation of the group Z of integers in the space C^2 defined by

$$\Pi(n) = \begin{pmatrix} 1 & n \\ 0 & 1 \end{pmatrix} \,, \qquad n \in Z$$

(it contains only one invariant subspace other than $\{0\}$ and C^2).

THEOREM III.1. (Schur's lemma). If (E, Π) is a finite dimensional irreducible representation, every linear operator in E which commutes with Π (i.e. with $\Pi(g)\ \forall g$) is a scalar operator.

PROOF. If T is such an operator, it has at least one eigenvalue λ, then the subspace $\mathrm{Ker}(T - \lambda I)$ is not zero and is invariant under Π ; hence it is equal to E, and $T = \lambda I$.

COROLLARY III.1. Every finite-dimensional irreducible representation of an abelian group G is one-dimensional, i.e. is a morphism of G into the group of non zero complex numbers.

THEOREM III.2. We assume G is finite.

(i) Every irreducible representation is finite-dimensional.

(ii) Every invariant vector subspace of a representation admits an invariant supplementary.

(iii) Every finite-dimensional representation is a direct sum of irreducible representations.

PROOF. (i) Let x be a non zero vector in E ; the vector subspace generated by the elements $\Pi(g)\cdot x$, $g \in G$, is invariant, hence equal to E ; on the other hand it is clearly finite-dimensional.

(ii) Let F be an invariant subspace ; it admits a supplementary F' ; let p be the projection from E onto F parallelly to F' ; set

$$F'' = \{x \in E \mid \sum_{g \in G} \Pi(g)\, p\, \Pi(g)^{-1} \cdot x = 0\} ;$$

then F'' is an invariant supplementary to F .

(iii) follows from (ii) by induction on dim E .

Continuous representations of topological groups.

DEFINITION III.4. Assume G is a topological group and E is a topological vector space. A representation Π of G in E is said to be continuous if the mapping $(g,x) \to \Pi(g)\cdot x$ is continuous from $G \times E$ into E .

One also says that Π is "strongly continuous" ; when E is finite-dimensional, this is equivalent to saying that $g \to \Pi(g)$ is continuous in the norm topology on Aut E for any norm on E ; by II.5. we know that if G is a Lie group every continuous finite-dimensional representation is C^{∞}.

III.2. Unitary representations of topological groups

DEFINITION III.5. A representation Π of a topological group G is said to be unitary if

a) E is a complex Hilbert space, with scalar product $(\,|\,)$

b) $\Pi(g)$ is a unitary operator for every $g \in G$, i.e.

$$(\Pi(g)\cdot x \mid \Pi(g)\cdot y) = (x|y) \qquad \forall x,y \in E .$$

c) Π is continuous, which is trivially equivalent to saying that $g \to \Pi(g)\cdot x$ is continuous for every x .

EXAMPLE. Take $G = R$, $E = L^2(R)$, and define two representations Π_1 and Π_2 of G in E as follows :

$$(\Pi_1(t)\cdot f)(x) = f(x+t)$$

$$(\Pi_2(t)\cdot f)(x) = e^{itx}\cdot f(x) \quad ;$$

conditions a) and b) above are trivially satisfied ; one sees rather easily using Lebesgue's theorem, that c) is satisfied for Π_2 ; hence it is also satisfied for Π_1 since Π_1 and Π_2 correspond to each other via Fourier transform. The representation Π_1 is called the regular representation of the group R ; it can be defined for an arbitrary locally compact group, replacing the Lebesgue measure dx by a Haar measure (but there is no longer Fourier transform!).

DEFINITION and PROPERTIES. Those of n^oIII.1. can be transposed as follows

- for the equivalence one demands that u be isometric,
- for a subrepresentation one assumes F is closed ,
- for the contragredient , one replaces the algebraic dual E^* by the topological dual E' .
- for the direct sum one replaces the "algebraic" direct sum of n^oIII.1 by the Hilbert sum,
- for the tensor product one replaces the "algebraic" tensor product of III.1. by the Hilbert tensor product.

One says Π is irreducible if there is no closed invariant subspace other than $\{0\}$ and E .

DEFINITION III.6. The set of all irreducible unitary representations of G will be denoted by $\hat{G}$.

THEOREM III.3. (Schur's lemma). If Π is irreducible, every continuous operator commuting with Π is scalar. In particular every irreducible unitary representation of an abelian group is one-dimensional, i.e. is a continuous morphism of G into the group of complex numbers of modulus 1 (this is called a character of G).

PROOF. Let T be a continuous operator commuting with Π ; write $T = T_1 + iT_2$ with T_1, T_2 hermitian ; then T_1 and T_2 commute with Π and we are led to the case of a hermitian T . Apply the spectral theorem : $T = \int \lambda d E_\lambda$; then E_λ commutes with Π, hence is equal to 0 or I ; there exists λ_o such that

$$E_\lambda = \begin{cases} 0 & \text{for } \lambda \leq \lambda_o \\ I & \text{for } \lambda \leq \lambda_o \end{cases}$$

then $T = \lambda_o \cdot I$.

THEOREM III.4. (i) Every closed invariant vector subspace admits an invariant closed supplementary.

(ii) Every finite-dimensional unitary representation is a direct sum of irreducible representations.

PROOF. (i) It is trivially seen that the orthogonal supplementary is invariant.

(ii) follows by induction on dim E .

REMARK III.2. Assertion (ii) is not true for infinite-dimensional representations : look at the regular representation of R !

THEOREM III.5. If G is compact, every irreducible unitary representation is finite dimensional.

PROOF. For $x,y,x',y' \in E$ set

$$\phi(x,x',y,y') = \int_G (\Pi(g)\cdot x|y)\cdot \overline{(\Pi(g)\cdot x'|y')}\, dg$$

where dg is the normalized Haar measure of G ; by the Fischer-Riesz theorem this can be written as

$$\phi(x,x',y,y') = (T_{y,y'}\cdot x|x')$$

where $T_{y,y'}$ is a continuous operator in E which commutes with Π ; by Schur's lemma $T_{y,y'}$ is a scalar $\psi(y,y')$; in the same way one proves that

$$\psi(y,y') = k\cdot(y|y')$$

for some $k \in C$; hence we can write

$$\int_G (\Pi(g)\cdot x|y)\cdot \overline{(\Pi(g)\cdot x'|y')}\, dg = k\cdot(x|x')\cdot(y|y') .$$

Let (e_i) be an orthogonal basis of E ; we have

$$1 = \int_G (\Pi(g)\cdot e_1|\Pi(g)\cdot e_1)\, dg$$

$$= \sum_i \int_G (\Pi(g)e_1|e_i)\cdot\overline{(\Pi(g)e_1|e_i)}\, dg$$

$$= \sum_i k\cdot(e_1|e_1)\cdot(e_i|e_i) = k\cdot \sum_i 1 \quad ;$$

this implies that the family (e_i) is finite.

THEOREM III.6. Suppose G is compact and (E,Π) is a continuous finite-dimensional representation of G . There exists a scalar product on E such that Π is unitary.

PROOF. Take an arbitrary scalar product $(\,|\,)_o$ on E and define a new scalar product $(\,|\,)$ by

$$(x|y) = \int (\Pi(g)\cdot x |\, \Pi(g)\cdot y)_o \; dg .$$

EXAMPLES. Theorems III.5. and III.6. apply to the Lie groups $O(n)$, $SO(n)$, $U(n)$, $SU(n)$, which are compact since their matrix coefficients are bounded ; in the next section we shall describe the representations of $SU(2)$ and $SO(3)$. For non compact groups (as $SL(2,R)$, $SL(2,C)$,...) the theory is much more difficult (see [6] for $SL(2,C)$).

III.3. Irreducible representations of $SU(2)$ and $SO(3)$. For every $s = 0, \frac{1}{2}, 1, \frac{3}{2}, 2, \ldots$ we denote by E^s the complex vector space of all polynomials of two complex variables u,v which are homogeneous and of degree $2s$; we define a representation Π of $SU(2)$ in E^s as follows

$$(\Pi(g)\cdot f)(u,v) = f\,(g^{-1}\cdot(u,v)) \;,$$

where $g\cdot(u,v)$ denotes the natural action of $SU(2)$ on C^2. Clearly Π is nothing but the $2s$-th symmetric power of the natural representation of $SU(2)$ in C^2.

Our space E^s has a basis consisting of the polynomials

$$f_m(u,v) = u^{s-m}\cdot v^{s+m} \;,$$

where $m = -s,-s+1,\ldots,s-1,s$, hence

$$\dim E^s = 2s+1 .$$

An easy computation shows that the infinitesimal representation is given by

$$(D\,\Pi)(X_1)\cdot f_m = \frac{i}{2}\,(s-m)\cdot f_{m+1} + \frac{i}{2}\,(s+m)\cdot f_{m-1}$$

$$(D\,\Pi)(X_2)\cdot f_m = \frac{1}{2}\,(s-m)\cdot f_{m+1} - \frac{1}{2}\,(s+m)\cdot f_{m-1}$$

$$(D\,\Pi)(X_3)\cdot f_m = -\,i\,m\cdot f_m .$$

THEOREM III.7. The representation Π of $SU(2)$ in E^s is irreducible.

PROOF. Let us introduce the operators U_+, U_-, U_3 in E^s :

$$U_{\pm} = -i\, D\,\Pi(X_1) \pm D\,\Pi(X_2)$$

$$U_3 = i\, D\,\Pi(X_3) \;;$$

we have

(III.1) $$U_{\pm} f_m = (s \mp m)\, f_{m \pm 1}$$

$$U_3\, f_m = m\, f_m \,.$$

Let F be a non-zero invariant vector subspace of E^s ; by Theorem III.6 we can consider the orthogonal projection P onto F (orthogonal for some invariant scalar product); P commutes with U_3 , which is diagonal with distinct diagonal entries, hence P itself is diagonal, and F contains at least one f_m ; formula (III.1) then shows that F contains all f_m's, hence is equal to E^s .

NOTATION. In what follows the representation Π of $SU(2)$ in E^s will be denoted by D_s ; we thus have

$$\dim D_s = 2s + 1 \quad .$$

One can prove that <u>these representations are all irreducible representations</u> of $SU(2)$ (see [3] or [6]).

<u>Particular cases</u>. D_0 is the trivial representation ; $D_{1/2}$ is the natural representation in C^2 ; D_1 is the complexified of the adjoint representation (one puts in correspondence f_{-1} and $X_1 - iX_2$, f_0 and $-X_3$, f_1 and $-X_1 - iX_2$) .

The representations D_s , realized in the spaces E^s , are called "spinor representations" and the elements of E^s are called "spinors".

REMARK III.3. A direct computation shows that

$$\sum_{j=1}^{3} (D\,\Pi(X_j))^2 = -s(s+1) \cdot I \quad ;$$

but the fact that the left member is a scalar operator follows from Schur's lemma and from the fact that $X_1^2 + X_2^2 + X_3^2$ belongs to the center of $\mathcal{U}(\mathfrak{g})$, if one considers X_j^2 as defined in $\mathcal{U}(\mathfrak{g})$.

THEOREM III.8. When s is an integer, the representation D_s defines, by passing to the quotient, an irreducible representation of

SO(3). In this way one obtains all irreducible representations of SO(3).

PROOF. Let us denote by T the morphism $SU(2) \to SO(3)$ constructed in II.5 ; for each representation ρ of $SO(3)$, $\rho \circ T$ is clearly a representation Π of $SU(2)$; moreover a representation Π of $SU(2)$ is of that form if and only if it satisfies $\Pi(-I) = I$; but this is the case if and only if s is an integer. Finally the correspondence $\rho \to \Pi$ preserves irreducibility.

REMARK III.4. When s is not an integer, some people still consider D_s as a (bivalued) representation of SO(3) ; actually it is preferable to consider it as a one-valued, but projective, representation, like those which we shall encounter in the case of the Galilean group.

III.4. Infinitesimal operators of unitary representations of Lie groups. Let us consider a unitary representation Π of a Lie group G in a Hilbert space $\mathcal{H}$. For every X in $\mathfrak{g}$ (Lie algebra of G), the operators $\Pi(\exp tX)$, $t \in R$, form a continuous one parameter group of unitary operators ; by Stone's theorem there exists a unique self-adjoint operator A_X in $\mathcal{H}$ such that

$$\Pi(\exp t X) = e^{itA_X} .$$

One is tempted to write $A_X = D\Pi(X)$, but it does not work because the domain of A_X depends on X, and one does not get a representation $D\Pi$ of $\mathfrak{g}$ in $\mathcal{H}$. It is possible to turn this difficulty, considering the subspace $\mathcal{H}^\infty$ of $\mathcal{H}$ consisting of all vectors $\xi \in \mathcal{H}$ such that the function $g \to \Pi(g)\xi$ is of class C^∞ and then defining

$$D\Pi(X)\xi = \frac{d}{dt} \Pi(\exp t X)\xi \Big|_{t=0} ;$$

one thus actually get a representation $D\Pi$ of $\mathfrak{g}$ in $\mathcal{H}^\infty$, but the space $\mathcal{H}^\infty$ is often not very easy to handle with.

Anyway, the selfadjoint operators A_X are important : in Quantum Mechanics they represent important observables of a system for which G is an invariance group.

EXAMPLE. Let us take $G = SO(3)$, $\mathcal{H} = L^2(R^3)$;

$$(\Pi(g)\psi)(p) = \psi(g^{-1} \cdot p), \quad g \in G, \psi \in \mathcal{H} ;$$

consider the basis elements y_1, y_2, y_3 of $\mathfrak{so}(3)$ introduced in II.5; one gets formally

$$A_{y_1} = p_3 \cdot \frac{\partial}{\partial p_2} - p_2 \cdot \frac{\partial}{\partial p_3}$$

$$A_{y_2} = p_1 \cdot \frac{\partial}{\partial p_3} - p_3 \cdot \frac{\partial}{\partial p_1}$$

$$A_{y_3} = p_2 \cdot \frac{\partial}{\partial p_1} - p_1 \cdot \frac{\partial}{\partial p_2} \quad .$$

The domain of A_{y_j} is the set of all $\psi \in H$ such that the limit

$$\lim_{t \to 0} \frac{1}{t} \left(\Pi(\exp t\, y_j)\, \psi - \psi \right)$$

exists in H. The space H^{∞} is the set of all ψ which are C^{∞} with respect to the angular variables. But one can also use other common domains for the A_{y_j}'s , for instance the Schwartz space $S(R^3)$.

IV. INDUCED REPRESENTATIONS. APPLICATION TO THE GALILEAN GROUP

IV.1. Induced representations. (For more details, see [5]). We consider a locally compact group G, a closed subgroup H and a unitary representation ρ of H in a Hilbert space K; we want to construct a representation of G.

Denote by X the coset space G/H; there is a natural action of G on X denoted by $g\cdot x$; it can be proved that there exists a quasi-invariant positive measure μ on X; this means that there exists a strictly positive function α on $G\times X$ such that

$$d\,\mu(g\cdot x) = \alpha(g,x)\,d\,\mu(x)$$

($\alpha(g,\cdot)$ is the Radon-Nikodym derivative of $d\,\mu(g\cdot)$ with respect to $d\,\mu(\cdot)$).

It can also be proved that there exists a Borel section s for $G\to X$, i.e. a Borel mapping such that

$$s(x)\cdot H = x \qquad \forall x\in X \quad ;$$

we define a mapping $\lambda: G\times X\to H$ by

$$\lambda(g,x) = s(x)^{-1}\cdot g\cdot s\,(g^{-1}\cdot x) \quad .$$

Finally we set $\mathcal{H}= L^2(X,\,\mu;\,K)$ and associate to each $g\in G$ the following operator $\Pi(g)$ in $\mathcal{H}$:

$$(\Pi(g)\cdot f)(x) = \alpha(g,g^{-1}\cdot x)^{-\frac{1}{2}}\cdot\rho(\lambda(g,x))\cdot f(g^{-1}\cdot x) \quad . \tag{IV.1}$$

It is easily checked that $\Pi(g)$ is a unitary operator, Π is a representation of G in $\mathcal{H}$ (the continuity is less evident and will be admitted), and that it is , up to equivalence, independent of the choice of μ and s (we recall that μ is unique up to equivalence).

DEFINITION IV.1. The representation Π constructed above is called the representation of G induced by ρ and denoted by $\mathrm{Ind}_H^G\,\rho$.

REMARK IV.1. One can give another realization of $\mathrm{Ind}_H^G\,\rho$, perhaps more natural but often less efficient : $\mathcal{H}$ is the space of all measurable functions $F: G\to K$ satisfying

(IV.2) $F(g\,h) = \rho(h)^{-1}\cdot F(g) \qquad g\in G\ ,\ h\in H$

and

$$\int_X \|F(g)\|^2\, d\,\mu(\dot g) < \infty$$

(this makes sense because $\|F(g)\|^2$ depends only on the class $\dot g$ of g by virtue of (IV.2)); Π is given by

$$(\Pi(g)\cdot F)(g') = \alpha(g, g^{-1}\cdot \dot g')^{-\frac{1}{2}}\cdot F(g^{-1}g')\ .$$

The relation between F and f is

$$f(x) = F(s(x))\ .$$

In that form the definition reminds that of induced representations of finite groups (but in that case $\alpha = 1$).

EXAMPLE. If $H = \{e\}$ and $K = \mathbb{C}$, Π is nothing but the regular representation of G.

IV.2. Application to semi-direct products. Here we assume G is a semi-direct product

$$G = B \rtimes A\ ;$$

this means that A and B are closed subgroups, B normal, and that every $g \in G$ can be written in a unique manner as $g = b\,a$. We moreover assume that B is finite-dimensional real vector space. We write $a\cdot b$ for $a\,b\,a^{-1}$ for $a\in A$, $b\in B$; this action of A on B is a real representation of A; we shall write (b,a) instead of $b\,a$, then the composition law becomes

$$(b,a)(b',a) = (b + a\cdot b', aa')\ .$$

We denote by B^* the dual vector space of B, and by $\langle b,x\rangle$ the value of $x\in B^*$ on $b\in B$. The group A acts naturally on B^* as follows:

$$\langle a\cdot x, b\rangle = \langle x, a^{-1}\cdot b\rangle\ .$$

We choose a point $x_o \in B^*$ and denote by S its stabilizer in A (also called "little group at x_o")

$$S = \{a\in A \mid a\cdot x_o = x_o\}\ ;$$

we then choose a unitary representation σ of S in a Hilbert space K;

we form the representation ρ of $B \ltimes S$ in K defined by

$$\rho(b,a) = e^{i<b,x_o>} \cdot \sigma(a), \quad b \in B, \; a \in S,$$

(check that it is indeed a representation!) ; finally we induce ρ to G and obtain a representation Π in a Hilbert space H which we shall describe more precisely.

We identify $X = G/B \ltimes S$ with the orbit $A \cdot x_o$ of x_o under A ; then the measure μ is a measure on $A \cdot x_o$; we choose a section $s: A \cdot x_o \to A$; then

$$\lambda((b,a),x) = (s(x)^{-1} \cdot b \; , \; \lambda(a,x)) \; ;$$

formula (IV.1) becomes

$$\text{(IV.3)} \qquad (\Pi\,(b,a)f)(x) = \alpha(a,a^{-1} \cdot x)^{-\frac{1}{2}} \cdot e^{i<b,x>} \cdot \sigma(\lambda(a,x)) \cdot f(a^{-1} \cdot x) \, .$$

One of the main results of Mackey's theory is the following

THEOREM IV.1. (i) The representation Π is irreducible if and only if σ is irreducible.

(ii) Two pairs $(x_o, \sigma), (x'_o, \sigma')$ lead to equivalent representations of G if and only if they are conjugated under A, i.e. if there exists $a \in A$ such that

a) $x'_o = a \cdot x_o$ (which implies $S' = a\,S\,a^{-1}$),

b) σ' is equivalent to the representation $s' \to \sigma\,(a^{-1}\,s'\,a)$.

(iii) If all the orbits of A in B^* are locally closed (=open in their closure), then the above procedure gives all irreducible unitary representations of G.

COROLLARY IV.1. If all the orbits of A in B^* are locally closed, one gets a complete parametrization of $\hat{G}$ by choosing one point in each orbit, and then an irreducible unitary representation of the stabilizer of this point in A.

EXAMPLES. It will be a useful exercise to write down the set $\hat{G}$ when G is the Euclidean group in R^2 or E^3.

REMARK IV.2. The Stone theorem mentioned in III.4. admits the following generalization: consider a unitary representation Π of some group R^m in a Hilbert space H ; the m basis elements of R^m give rise to m self-adjoint commuting operators $T_1,\dots,T_m$ in H ; by the spectral theorem for commuting families of self-adjoint operators there exists a spectral measure E on R^n with values in the projectors of H, such that

$$T_j = \int_{R^m} \lambda_j \, d\, E_\lambda$$

(we have written $\lambda = (\lambda_1, \dots, \lambda_n)$). Then

$$\Pi(t) = \int_{R^m} e^{i\langle t,\lambda\rangle} \, d\, E_\lambda$$

where $\langle t,\lambda\rangle = t_1\lambda_1 + \dots + t_m\lambda_m$; one can say that $\Pi(t)$ is the operator of multiplication by the function $\lambda \to e^{i\langle t,\lambda\rangle}$. In the situation of Theorem IV.1, the spectral measure associates with each Borel subset Y of B* the projection onto the subspace $L^2(Y \cap A\cdot x_o, \mu; K)$; in particular this spectral measure is concentrated on the orbit $A \cdot x_o$.

IV.3. Representations of the Galilean group. An element of the Galilean group G(see §I) will be denoted by

$$g = (A,B,C,D)$$

with $A \in SU(2)$, $B,C \in R^3$, $D \in R$; the composition law is

$$(A,B,C,D)(A',B',C',D') = (AA', AB'+B, AC'+C+BD', D+D')$$

(that is, we have replaced SO(3) by its covering SU(2) ; it acts on R^3 via the natural action of SO(3)). Accordingly to what has been said in §I about the Lie algebra of G , we must consider projective representations, i.e.mappings Π from G into the unitary group of some Hilbert space H satisfying

$$\Pi(g)\cdot\Pi(g') = \sigma(g,g')\cdot\Pi(gg')$$

where σ is a Borel function on $G\times G$ satisfying

$$|\sigma(g,g')| = 1$$

$$\text{(IV.4)} \qquad \sigma(g',g'')\cdot\sigma(g\,g',g'')^{-1}\cdot\sigma(g,g'g'')\cdot\sigma(g,g')^{-1} = 1$$

(such a function is called a "2-cocycle" or a "Mackey multiplier"). Trying to "integrate" the projective representation of $\mathfrak{g}$ given in § I , one can convince himself that a good candidate for σ is

(IV.5) $$\sigma(g,g') = \exp i\, m(B.AC' + \frac{1}{2}\ |B|^2\ D')$$

where m is a fixed non zero real number. Now, like in § I , one can replace the projective representations of G by the ordinary representations of some other group $\tilde{G}$: $\tilde{G}$ is the set $G \times T$ (T = the group of complex numbers of modulus 1) with the following composition law :

$$(g,E)\cdot(g',E') = (gg',\ E\cdot E'\cdot\ \sigma(g,g'))$$

(the associativity follows from (IV.4)). Now the projective representations of G with multiplier σ correspond bijectively to the representations Π of $\tilde{G}$ satisfying

(IV.6) $$\Pi(1,E) = E$$

by the correspondence

(IV.7) $$\Pi(g) = \Pi(g,1)\ .$$

We are then led to the following

Problem 1. Find all irreducible unitary representations Π of $\tilde{G}$ satisfying (IV.5).

Actually it will be convenient to make a slight change : to replace T by R , σ by τ with

$$\tau(g,g') = m(B\cdot AC' + \frac{1}{2}\ |B|^2\ D'\ )$$

and $\tilde{G}$ by $\overline{G} = G \times R$ with the composition law

$$(g,E)\cdot(g',E') = (gg',E + E' + \tau(g,g'))\ .$$

then (IV.6) and (IV.7) become respectively

(IV.8) $$\Pi(1,E) = e^{iE}\ ,$$

(IV.9) $$\Pi(g) = \Pi(g,0)\ .$$

We are now faced to

Problem 2. Find all irreducible unitary representations Π of $\overline{G}$ satisfying (IV.8).

We shall apply IV.2, the subgroups A and B being replaced respectively by

$$\overline{G}_o = \{a = (A,B,0,0,0)\}$$
$$\overline{G}_1 = \{b = (I,0,C,D,E)\} \sim R^5 \quad ;$$

the action of $\overline{G}_o$ on $\overline{G}_1$ by inner automorphisms is given by

$$(A,B)\cdot(C,D,E) = (AC+BD, D, E+\frac{m}{2}(2\,AC\cdot B+D\,|B|^2)) \;.$$

The elements of $\overline{G}_1^*$ will be denoted by

$$x = (p, p_o, w)$$

where $p \in R^3, p_o$ and $w \in R$, with the duality

$$<b,x> = C\cdot p + D\,p_o + E\,w \;.$$

The action of $\overline{G}_o$ on $\overline{G}_1^*$ is given by

$$(A,B)\cdot(p,p_o,w) = (Ap - mwB \;,\; p_o - Ap\cdot B + \frac{mw}{2}\,|B|^2 \;,\; w) \;.$$

The orbits are defined by the equations

$$w = \alpha$$

$$\alpha\,p_o - \frac{1}{2m}\,|p|^2 = \beta$$

where α and β are arbitrary real constants ; we denote this orbit by $\mathcal{O}_{\alpha\beta}$. Actually the orbits with $\alpha = 0$ would lead to representations not satisfying (IV.8), so that we shall suppose $\alpha \neq 0$. As a point x_o in $\mathcal{O}_{\alpha\beta}$ we choose

$$x_o = (0, \frac{\beta}{\alpha}, \alpha) \quad ;$$

its stabilizer S in $\overline{G}_o$ is

$$S = \{(A,0)\} \sim SU(2) \;.$$

As a section $\mathcal{O}_{\alpha\beta} \to \overline{G}_o$ we choose

$$(p,p_o,\alpha) \to (I, -\frac{p}{m\alpha}) \quad ;$$

then the function $\lambda(a,x)$ of section II.2. becomes

$$\lambda((A,B),(p,p_o,\alpha)) = (A,0) \;.$$

The orbit $\mathcal{O}_{\alpha\beta}$ admits an invariant measure (and not only a quasi-invariant one), namely the Lebesgue measure $d\,p$. We have now to choose an irreducible representation σ of S ; following III.3, σ is of the form $D_s, s = 0, \frac{1}{2}, 1, \frac{3}{2}, \ldots,$ acting on E^s . The corresponding representation Π of $\bar{G}$ acts on the space $H = L^2(\mathcal{O}_{\alpha,\beta}, dp ; E^s) = L^2(R^3, dp, E^s)$ and formula (IV.3) becomes

$$(\text{IV.10}) \qquad (\Pi(A,B,C,D,E)f)(p) =$$

$$= \exp i\,(C\cdot p + \frac{1}{2m\alpha} D|p|^2 + \frac{D\beta}{\alpha} + E\alpha)\cdot D_s(A)\cdot f(A^{-1}p + m\,A^{-1}B) .$$

In order to satisfy (IV.8) we must take $\alpha = 1$; moreover if we change β the representation will change only by a scalar multiple $\exp i D\beta$, hence will remain projectively equivalent to the former; hence we can take $\beta = 0$, so that formulae (IV.10) and (IV.9) give

$$(\text{IV.11}) \qquad (\Pi(A,B,C,D)\, f)(p) =$$

$$= \exp i\,(C\cdot p + \frac{1}{2m} D|p|^2)\cdot D_s(A)\cdot f(A^{-1}p + m\,A^{-1}B) .$$

We have thus proved the following

THEOREM IV.2. For every $s \in \frac{1}{2}N$ and every $m \in R\setminus\{0\}$, formula (IV.11) defines an irreducible unitary projective representation of G in the Hilbert space $L^2(R^3; E^s)$, with multiplier σ given by (IV.5) ; moreover this construction yields, up to a scalar multiple, all irreducible projective representations with multiplier σ .

Interpretation of the above results. First we indicate the infinitesimal operators of Π, but writing H as $L^2(R^3)\otimes E^s$ instead of $L^2(R^3; E^s)$, and using $\hat{Q}_j, \hat{P}_j, \hat{L}_j, \hat{H}$ as defined in Remark I.3:

$$D\Pi(\ell_j) = -\frac{i}{\hbar}\hat{L}_j\otimes I + I\otimes D\,D_s(\ell_j) ,$$
$$D\Pi(q_j) = -i\,m\,\hat{Q}_j\otimes I ,$$
$$D\Pi(p_j) = \frac{i}{\hbar}\hat{P}_j\otimes I ,$$
$$D\Pi(h) = \frac{i}{\hbar^2}\hat{H}\otimes I ;$$

thus we obtain what we denoted by $U(\ell_j)$,... in §I with one difference: we have replaced $L^2(R^3)$ by $L^2(R^3)\otimes E^s$ and added to $U(\ell_j)$ the additional term $I\otimes D\,D_s(\ell_j)$. The physical interpretation of this change is that here we study a particle having <u>spin</u> s (and in § I a "spinless" or "scalar" particle) and the 3 operators $I\otimes D\,D_s(\ell_j)$

represent the 3 components of the spin (or intrinsic) kinetic momentum.

On the other hand one can look at H as a space of distributions on R^4 supported by the paraboloid having equation

$$p_o - \frac{1}{2m} |p|^2 = 0 \quad ;$$

if we perform a Fourier transform on 4 variables and remember that the variable conjugate to p_o is $-\hbar t$ we get

$$\frac{\partial \psi}{\partial t} = \frac{i\hbar}{2m} \Delta \psi$$

which is the Schrödinger equation for our free particle.

IV.4. Concluding remarks.

REMARK IV.3. The group $\overline{G}$ contains as a subgroup the set of elements $(I,B,C,0,E)$ which we shall write (B,C,E) ; the composition law is

$$(B,C,E) \cdot (B',C'E') = (B + B',C + C', E + E' + m\, B\cdot C') \quad ;$$

this group is called the Heisenberg group in 3 dimensions.

By formula (IV.10) we have

$$(\Pi(B,C,E)f)(p) = \exp i \ (C \cdot p+E) \cdot f(p+m\cdot B) \quad ;$$

one can show that if $s = 0$, this representation is irreducible; moreover, there is a theorem by Stone-von Neumann asserting that conversely every irreducible representation Π of this group, satisfying $\Pi(0,0,E) = \exp i\,E$, is equivalent to ours. This result was the starting point of Kirillov's theory (also called "orbit method") for the representations of nilpotent Lie groups : for a simply connected nilpotent Lie group G , it states a bijective correspondence between $\hat{G}$ and the orbits of G acting on g^* (= dual vector space of g) via the contragredient of the adjoint representation.

This theory was later generalized (but the results are not so simple) to a large class of solvable Lie groups by Bernat and Auslander-Kostant.

Perhaps it will be useful to say that a Lie group G is called solvable (resp. nilpotent) if it contains closed subgroups

$$G = G_o \supset G_1 \supset \dots \supset G_n = \{1\}$$

such that G_{i+1} is normal in G_i (resp. in G) and G_i/G_{i+1} is abelian for all $i = 0,\dots,n-1$.

At the opposite side we have the class of semi-simple Lie groups, like $SL(n,k)$, $SO(n,k)$, $SU(n)$, etc. defined by the condition that they contain no non-discrete closed normal subgroups ; for these groups the representation theory is thoroughly different; the compact case was already known by E.Cartan and H.Weyl ; the non-compact case began with $SL(2,R)$, $SL(2,C)$ (Bargmann, Gelfand-Naimark) and became a general theory with the works of Harish-Chandra, Langlands,etc.

Finally, Mackey's theory, partially explained in IV.2, theoretically allows to reduce the case of arbitrary Lie groups to that of semi-simple ones (in fact every Lie group contains a largest solvable normal subgroup, and the quotient is semi-simple); Duflo recently made this reduction much more explicit.

REMARK IV.4. Let us briefly examine what happens if one replaces the Quantum Mechanics considered in §I by the Relativistic Quantum Mechanics. First one has to replace the Galilean group by the Poincaré group G which is defined as follows: denote by G_o' the Lorentz group, i.e. the set of all linear automorphisms of R^4 preserving the quadratic form $x_o^2-x_1^2-x_2^2-x_3^2$; its neutral connected component G_o is the set of elements of G_o' having determinant 1 and preserving the future light cone ($x_o^2-x_1^2-x_2^2-x_3^2 = 0, x_o > 0$) ; then G is the semi-direct product of G_o with $G_1 = R^4$ for the natural action of G_o on G_1 . Mackey's theory applies specially succesfully to the orbits which are parts of two-sheeted hyperboloids (then the stabilizer S is $SO(3)$) or the light cone (then S is the euclidean group in two dimensions). In the case of the superior part of an hyperboloid $p_o^2-p_1^2-p_2^2-p_3^2 = m^2$, $m > 0$, one interprets m as the mass of the particle, moreover the choice of a representation of $SO(3)$ introduces an integer s still interpreted as the spin (to obtain non integral spins one has to replace G by a two-sheeted covering). Finally the equation of the hyperboloid leads by Fourier Transform, no longer to the Schrödinger equation, but to the Klein-Gordon equation.

REFERENCES

[1] J.P.Serre, Lie algebras and Lie groups, Benjamin, 1965.

[2] G.Hochschild, The structure of Lie groups, Holden-Day, 1965.

[3] J.P.Serre, Algèbres de Lie semi-simple complexes, Benjamin, 1966.

[4] M.Naimark-A.Stern, Théorie des représentations des groupes, Editions de Moscou, 1979.

[5] A.Kirillov, Eléments de la théorie des représentations, Editions de Moscou, 1974.

[6] M.Naimark, Linear representations of the Lorentz group.

[7] V.S.Varadarajan, The geometry of Quantum Theory, Van Nostrand, 1970 .

PROBABILITY AND GEOMETRY OF BANACH SPACES

J.Hoffmann-Jørgensen

1. INTRODUCTION

Ever since the dawning of probability two results have been in the center of research and study. The two results, also called the two pearls of probability, are the law of large numbers and the central limit theorem.

The law of large numbers was discovered in 1695 by James Bernoulli (Published in his book Ars Conjectandi in 1713), only 40 years after the foundation of probability (1654). Loosely speaking the law of large numbers states that the averages of random variables with the same mean μ converges to μ, provided that the random variables do not depend too much of each other. I.e. that

$$(1.1) \qquad \frac{1}{n} \sum_{j=1}^{n} X_j \to \mu \qquad \text{as } n \to \infty$$

if $E X_j = \mu$ for all j and certain independence conditions are satisfied. There are several convergence notions possible in (1.1). If the convergence is convergence in probability or in L^p we speak about a weak law, and if the convergence is a.s. (=almost sure) convergence, we speak about a strong law. Let me state 3 strong laws, which we shall extend to Banach spaces (see § 5):

(1.2) If $X_1, X_2, \ldots$ are independent, integrable and identically distributed, then (1.1) holds a.s.

(1.3) If $X_1, X_2, \ldots$ are independent, with the same means and $\sup_n \mathrm{Var}\, X_n < \infty$, then (1.1) holds a.s.

(1.4) If $X_1, X_2, \ldots$ are independent with mean μ and $\sum n^{-p} E|X_n - \mu|^p < \infty$ for some $1 \leq p \leq 2$, then (1.1) holds a.s.

(1.2) is the classical law of large numbers and it goes back to

Borel, Kolmogorov and Khinchine. (1.3) and (1.4) with $p=2$ are due to Kolmogorov and (1.4) with $1 \leq p < 2$ is due to Chung. We shall later see that (1.2) holds in any Banach space, (1.3) holds in B-convex Banach spaces, and (1.4) holds in Banach spaces of type p.

The law of large numbers gives information about the order of magnitude of partial sums of independent random variables. However a much more detailed information of the partial sums is provided by the central limit theorem. The central limit theorem was discovered by Abraham de Moivre in 1733 (Doctrine of Chances, 2 nd edition) and it states that if we normalize the sum with $\frac{1}{\sqrt{n}}$ instead of $\frac{1}{n}$, then the limit of the accompanying laws is gaussian. Again the central limit theorem is not just one theorem but a huge complex of theorems, and together with the law of large numbers it has been subjected to an intensive study in the last 250 years. We shall here only consider the most simple one, viz :

(1.5) If $X_1, X_2, \ldots$ are independent, identically distributed with mean μ and variance σ^2, then for all $x \in R$ we have

$$P\left(\frac{1}{\sqrt{n}} \sum_{j=1}^{n} (X_j - \mu) \leq x\right) \underset{n\to\infty}{\longrightarrow} \frac{1}{\sigma\sqrt{2\pi}} \int_{-\infty}^{x} e^{-t^2/2\sigma^2}\, dt$$

Note that the limit is nothing but the <u>gaussian</u> (or <u>normal</u>) distribution function with mean 0 and variance σ^2, which we shall denote by $N(0, \sigma^2)$. It is well known that the limit statement above is equivalent to

$$\mathcal{L}\left(\frac{1}{\sqrt{n}} \sum_{j=1}^{n} (X_j - \mu)\right) \to N(0, \sigma^2)$$

where $\mathcal{L}(X)$ is the distribution law of X and $\to$ denotes weak convergence of probability measures.

(1.5) is due to Chebysev and his two pupils, Markov and Lyapounov. We shall see later that (1.5) holds in Banach spaces of type 2, of course with a suitable definition of gaussian measure on Banach spaces (see Theorem 6.1)

It is probably by now clear that the main theme of these lectures is the study of sums of independent random variables:

$$S_n = \sum_{j=1}^{n} X_j$$

We shall actually begin our investigations with the study of the convergence of S_n. It turns out that loosely speaking any kind of convergence or even boundedness, no matter how weak, implies the strongest possible convergence, viz. a.s. convergence. To be precise, if $\{X_n\}$ is an independent sequence of random variables then the following 2 statements are equivalent:

(1.6) $\{S_n \mid n \geq 1\}$ is stochastically bounded (i.e. bounded in $L^o(\Omega, F, P)$)

(1.7) $\exists a_n \in R$ so that $\lim(S_n - a_n)$ exists a.s. and $\sup |a_n| < \infty$

And if (1.6) holds, then (1.7) holds with $a_n \equiv 0$ in each of the following 4 cases:

(1.8) $\{\mathcal{L}(S_n)\}$ is weakly convergent

(1.9) X_n is even $n \geq 1$ (i.e. $\mathcal{L}(X_n) = \mathcal{L}(-X_n)\ \forall n$)

(1.10) $E X_n = 0$ and $\sup_n E|S_n| < \infty$

(1.11) $E X_n = 0$ and $E(\sup_n |X_n|) < \infty$

This result, which is called <u>the equivalence theorem</u>, holds partially in arbitrary Banach spaces (see Theorem 4.1), and it holds fully in Banach spaces not containing c_o (see Theorem 4.3).

2. PREREQUISITES

Let (Ω, F, P) be a probability space which we will fix once for all in all of these lectures. As usual in probability we shall tend to forget about the underlying probability space (Ω, F, P) and concentrate about the study of random variables, random vectors and stochastic processes. We shall allow ourselves to think of (Ω, F, P) being so large, that we can define any set of random variables or vectors we like. This attitude is justified by Kolmogorov's consistency theorem, which loosely speaking states, that any set of random variables or vectors exists, unless there are obvious reasons for it not to exist. I shall in this section fix the notation and briefly describe the results from the general probability theory which is needed in the sequel.

I. Random variables (1): Let (S, B) be a measurable space (i.e. a set S and B is a σ-algebra on S). Then an S-valued random variable is a measurable map X from (Ω, F) into (S, B). An S-valued stochastic process with time set Θ, is an index family $X = \{X(\theta) \mid \theta \in \Theta\}$ of S-valued random variables.

(2): Let S be a set and H is a set of real valued functions on S, then $\sigma(H)$ denotes the smallest σ-algebra on S making all functions in H measurable. And $\sigma(S, H)$ denotes the weakest topology on S making all functions in H continuous.

(3): Let S be a topological space, then $C(S)$ denotes the set of all real valued continuous functions on S, and $C(S)$ denotes the set of all bounded continuous real valued functions on S. We shall always consider a topological as a measurable space with respect to its Borel σ-algebra: $B(S)$ = the smallest σ-algebra containing all open subsets of S. Thus an S-valued random variable is a Borel measurable map from (Ω, F) into $(S, B(S))$. But occasionally we shall meet the Baire σ-algebra, $Ba(S) = \sigma(C(S))$. Note that $Ba(S) = B(S)$, if S is metrizable.

(4): If S is a set, then $B(S)$ denotes the set of all bounded real valued functions on S. And then $B(S)$ is a Banach space under the sup-norm :

(2.1) $$||f||_\infty = \sup_{s\in S} |f(s)| \qquad \forall f \in B(S)$$

If S is a topological space, then $C(S)$ is a $||\ ||_\infty$-closed linear subspace of $B(S)$.

(5): If X is a real random variable, then the <u>mean</u> or <u>expectation</u> of X, denoted $E X$, is simply the integral

(2.2) $$E X = \int_\Omega X \, d P$$

provided of course, that the integral exists.

(6): Let $(S, \mathcal{B})$ be a measurable space, and X an S-valued random variable. Then <u>the law</u> of X, or <u>the distribution law</u> of X, is the image measure of P under X, and it is denoted $\mathcal{L}(X)$. I.e. if $\mu = \mathcal{L}(X)$ then

(2.3) $$\mu(B) = P(X \in B) \qquad \forall B \in \mathcal{B}.$$

Note that μ is a probability measure on $(S, \mathcal{B})$. If $\mu = \mathcal{L}(X)$, then we write $X \sim \mu$, and if $\mathcal{L}(X) = \mathcal{L}(Y)$ we write $X \sim Y$, and say, that Y is a <u>copy</u> or a <u>version</u> of X, and that X and Y are <u>identically distributed</u>. Note that if $X \sim Y$, then

(2.4) $$P(X \in B) = P(Y \in B) \qquad \forall B \in \mathcal{B}$$

(2.5) $$E\,\phi(X) = E\,\phi(Y) \qquad \forall \phi \text{ measurable: } S \to \overline{R}_+ .$$

This means that from a probabilistic viewpoint X and Y are identical objects, even though they may be very different as functions (they could even be defined on two different probability spaces).

(7): Let $(S_t, \mathcal{B}_t)$ be a measurable space and X_t an S_t-valued random variable for all $t \in T$. Then $\{ X_t | t \in T \}$ is said to be <u>independent</u> if

(2.6) $$P(X_t \in B_t \quad \forall t \in \alpha) = \prod_{t\in\alpha} P(X_t \in B_t)$$

for all finite sets $\alpha \subseteq T$ and all $\{ B_t | t \in \alpha\}$ with $B_t \in \mathcal{B}_t$ $\forall t \in \alpha$. Suppose that $\{X_t | t \in T\}$ are independent, and let $\{ T(u) | u \in U\}$ be mutually disjoint subsets of T, put

$$S_u = \prod_{t\in T(u)} S_t \,,\quad \mathcal{B}_u = \bigotimes_{t\in T(u)} \mathcal{B}_t \quad (=\text{the product } \sigma\text{-algebra})$$

If ϕ_u is a measurable map from $(S_u, \mathcal{B}_u)$ into a measurable space (M_u, A_u) for all $u \in U$, and $Y_u = \phi_u((X_t)_{t\in T(u)})$, then $\{Y_u \mid u \in U\}$ are independent.

(8): Let $X_1,\ldots,X_n$ be independent random variables taking values in $(S_1, \mathcal{B}_1),\ldots,(S_n, \mathcal{B}_n)$. Consider <u>the product space</u> $(S,\mathcal{B})$:

$$S = S_1 \times \ldots \times S_n \,,\quad \mathcal{B} = \mathcal{B}_1 \otimes \ldots \otimes \mathcal{B}_n$$

Let $\mu_j = \mathcal{L}(X_j)$, then $X = (X_1,\ldots,X_n)$ is an S-valued random variable, such that $\mu = \mathcal{L}(X)$ is <u>the product measure</u> : $\mu_1 \otimes \ldots \otimes \mu_n$. I.e.

(2.7) $$\mathcal{L}((X_1,\ldots,X_n)) = \mathcal{L}(X_1) \otimes \ldots \otimes \mathcal{L}(X_n)$$

provided that $X_1,\ldots,X_n$ are independent. Hence if ϕ_j is measurable: $S_j \to R$ and $E|\phi_j(X_j)| < \infty$, then

(2.8) $$E\{\prod_{j=1}^{n} \phi_j(X_j)\} = \prod_{j=1}^{n} E\{\phi_j(X_j)\}$$

by Fubini's theorem.

(9): Suppose that X and Y are independent with values in $(S_1, \mathcal{B}_1)$ and $(S_2, \mathcal{B}_2)$. If $\phi: S_1 \times S_2 \to R$ is measurable and $E|\phi(X,Y)| < \infty$ then

(2.9) $$E\,\phi(X,Y) = E\,\psi(Y) \quad \text{where} \quad \psi(s) = E\,\phi(X,s)$$

II. <u>Convergence in law</u> (1): Let S be a topological space, then $ca_\tau^+(S)$ denotes the set of all <u>positive finite measures</u> μ on $(S, \mathcal{B}(S))$ which are τ-<u>smooth</u>, i.e.

(2.10) $$\int_S (\sup_i f_i)\, d\mu = \sup_i \int_S f_i \, d\mu$$

whenever $\{f_i \mid i \in I\}$ is an upwards filtering family of non-negative lower semicontinuous functions on S. And $Rad^+(S)$ denote the set of all finite positive Radon measures on S, i.e. all positive finite measures μ on $(S, \mathcal{B}(S))$ such that for all $B \in \mathcal{B}(S)$ we have

(2.11) $$\mu(B) = \sup\{\mu(K) \mid K \text{ closed, compact and } K \subseteq B\}$$

The vector space $ca_\tau(S)$ (respectively $Rad(S)$) is the set of all real valued measures μ on $(S, \mathcal{B}(S))$ whose total variation $|\mu|$ (see [11, III.1.4]) belongs to $ca_\tau^+(S)$ (respectively $Rad^+(S)$).

(2): $Rad(S) \subseteq ca_\tau(S)$. And if S is metrizable then $\mu \in ca_\tau(S)$ if and only if $|\mu|(S \setminus S_o) = 0$ for some closed separable set S_o in S. If S is a complete metric space, then $Rad(S) = ca_\tau(S)$.

(3): On $ca_\tau(S)$ we define the <u>weak topology</u> to be the weakest topology making the functions

$$\mu \to \int_S f \, d\mu : ca_\tau(S) \to R$$

continuous for all $f \in C(S)$. If a net $\{\mu_\alpha\}$ in $ca_\tau(S)$ converges weakly to $\mu \in ca_\tau(S)$, we write $\mu_\alpha \overset{\sim}{\to} \mu$. I.e.

(2.12) $$\mu_\alpha \overset{\sim}{\to} \mu \iff \lim_\alpha \int_S f \, d\mu_\alpha = \int_S f \, d\mu \quad \forall f \in C(S)$$

(4): A finite positive measure μ on $(S, \mathcal{B}(S))$ is said to be <u>tight</u>, if for all $\varepsilon > 0$ there exists a compact closed set $K \subseteq S$, so that $\mu(S \setminus K) \leqq \varepsilon$. It is easily checked, that a finite positive measure μ on $(S, \mathcal{B}(S))$ is tight and τ-smooth, if and only if μ is a Radon measure. A set M $ca_\tau^+(S)$ is said to be <u>uniformly tight</u> if $\{\mu(S) \mid \mu \in M\}$ is bounded and for all $\varepsilon > 0$ there exists a compact closed set $K \subseteq S$, so that $\mu(S \setminus K) \leqq \varepsilon \quad \forall \mu \in M$.

(5): It is easily checked that, every uniformly tight set $M \subseteq ca_\tau^+(S)$ is relatively weakly compact. The converse is true provided that S is a complete metric space (<u>Prohorov's theorem</u>). The completeness is indispensable in Prohorov's theorem, but if S is metrizable, then every countable compact subset of $Rad^+(S)$ is uniformly tight. In particular every weakly convergent sequence of positive Radon measures on a metrizable space is uniformly tight.

(6): Let (S, ρ) be a pseudometric space. Then $Lip(S, \rho)$ denotes the set of all <u>bounded</u>, <u>Lipschitz' functions</u>, $f: S \to R$ (i.e. for some $K > 0$ we have $|f(s)| \leqq K$ and $|f(s) - f(t)| \leqq K \cdot \rho(s,t) \ \forall s,t$). Then $Lip(S, \rho)$

is a Banach space under the norm:

(2.13) $$\|f\|_\rho = \|f\|_\infty + \sup_{\rho(s,t)>0} \left\{ \frac{|f(s)-f(t)|}{\rho(s,t)} \right\} .$$

$M(s,\rho)$ denotes the dual space of $(Lip(S,\rho), \|\cdot\|_\rho)$ with its <u>dual norm</u> :

$$\|\xi\|_\rho = \sup \{ |<\xi,f>| : f \in Lip(S,\rho), \|f\|_\rho \leq 1 \}$$

Note that $ca_\tau(S)$ may be considered a subset of $M(S,\rho)$ by the identification:

$$<\mu,f> = \int_S f \, d\mu$$

so we have the $\|\cdot\|_\rho$-norm on $ca_\tau(S)$:

$$\|\mu\|_\rho = \sup\{ |\int f \, d\mu| : f \in Lip(S,\rho), \|f\|_\rho \leq 1\}$$

(7): Let (S,ρ) be a pseudometric space. It can then be shown, that the $\|\cdot\|_\rho$-topology on $ca_\tau^+(S)$ (but <u>not</u> in general on $ca_\tau(S)$) equals the weak topology. I.e. if $\{\mu_\alpha\}$ is a net in $ca_\tau^+(S)$ then

(2.14) $$\lim_\alpha \|\mu_\alpha - \mu\|_\rho = 0 \iff \mu_\alpha \overset{\sim}{\to} \mu$$

Thus $ca_\tau^+(S)$ with its weak topology is metrizable under the metric $\|\cdot\|_\rho$, which is a norm on $ca_\tau(S)$. Usually one considers the so called <u>Prohorov metric</u> : $\bar\rho$, rather than $\|\cdot\|_\rho$, where

$$\bar\rho(\mu,\nu) = \inf\{\alpha > 0 | \mu(B) \leq \alpha + \nu(B^\alpha) \ \forall B \in \mathcal{B}(S)\} \text{ if } \mu(S) \geq \nu(S)$$

where $B^\alpha = \{ s | \exists t \in B : \rho(s,t) < \alpha\}$. Then $\bar\rho$ is a metric and

(2.15) $$\tfrac{1}{3}\|\mu-\nu\|_\rho \leq \bar\rho(\mu,\nu) \leq 2\sqrt{\|\mu-\nu\|_\rho} \qquad \forall \mu,\nu \in ca_\tau^+(S) : \mu(S),\nu(S) \leq 1.$$

So $\bar\rho$ and $\|\cdot\|_\rho$ are equivalent metrics on $ca_\tau^+(S)$ (see[10]).

(8): Let (S,ρ) be a pseudometric space. If (S,ρ) is separable, analytic or complete, then so is $ca_\tau^+(S)$ under $\|\cdot\|_\rho$ and under $\bar\rho$. Thus if (S,ρ) is complete then $ca_\tau^+(S)$ is a $\|\cdot\|_\rho$-closed convex cone in $M(S,\rho)$. But $ca_\tau(S)$ is only $\|\cdot\|_\rho$-closed if ρ is a <u>discrete pseudometric</u> (i.e. $\exists \alpha>0$ so that $\rho(s,t) \geq \alpha$ $\forall (s,t)$ with $\rho(s,t) > 0$).

(9): Let S be a set and $H \subseteq B(S)$, such that $f \cdot g \in H$ for all $f, g \in H$. Let S be equipped with its $\sigma(S,H)$-topology, and let $\{\mu_\alpha\}$ be a net in $ca^+_\tau(S)$ and $\mu \in ca^+_\tau(S)$. It can then be shown that we have:

$$(2.16) \qquad \mu_\alpha \overset{\sim}{\to} \mu \Longleftrightarrow \lim_\alpha \int_S f \, d\mu_\alpha = \int_S f \, d\mu \qquad \forall f \in H.$$

(10): Prob(S) denotes the set of all <u>probability measures</u> $\mu \in ca^+_\tau(S)$, i.e. $\mu(S) = 1$. Then Prob(S) is a weakly closed convex subset of $ca_\pi(S)$. If $\{X_n\}$ is a sequence of S-valued random variables with $\mu_n = \mathcal{L}(X_n) \in \text{Prob}(S)$, we say that $\{X_n\}$ <u>converges in law</u> to μ_o or to X_o, and we write $X_n \overset{\sim}{\to} \mu_o$ or $X_n \overset{\sim}{\to} X_o$, if $\mu_n \overset{\sim}{\to} \mu_o$. Hence

$$(2.17) \qquad X_n \overset{\sim}{\to} X_o \Longleftrightarrow \lim_{n\to\infty} E\, f(X_n) = E\, f(X_o) \qquad \forall f \in C(S)$$

$$(2.18) \qquad X_n \overset{\sim}{\to} \mu_o \Longleftrightarrow \lim_{n\to\infty} E\, f(X_n) = \int_S f \, d\mu_o \qquad \forall f \in C(S)$$

Note that $X_n \overset{\sim}{\to} X_o$ by no means determines X_o uniquely, but it determines the law of X_o uniquely, provided S is completely regular.

III. <u>Measurable linear spaces</u> (1) : Let E be a linear space and $\mathcal{B}$ a σ-algebra on E. Then we say, that $(E, \mathcal{B})$ is a <u>measurable linear space</u>, if <u>addition</u> : $(x,y) \to x+y$ and <u>scalar multiplication</u> : $(x,t) \to t\,x$ are measurable, when $E \times E$ and $E \times R$ have their product σ-algebras $\mathcal{B} \otimes \mathcal{B}$ and $\mathcal{B} \otimes \mathcal{B}(R)$.

(2): Let $(E, \mathcal{B})$ be a measurable linear space and X an E-valued random variable, then we say that X is <u>even</u> if $X \sim (-X)$. Let $(E_n, \mathcal{B}_n')$ be a measurable linear space and X_n an E_n-valued random variable for $n = 1,2,\ldots$. Put

$$E^\infty = \prod_{n=1}^{\infty} E_n \qquad \text{and} \qquad \mathcal{B}^\infty = \bigotimes_{n=1}^{\infty} \mathcal{B}_n$$

Then $X = (X_1, X_2, \ldots) = (X_n)$ is an E^∞-valued random variable, and we say that the sequence (X_n) is <u>symmetric</u> if $(X_n) \sim (\pm X_n)$ for all choices of signs $\pm$, i.e. if the law of (X_n) on E^∞ equals the law of $(\alpha_n X_n)$ on E^∞ for all sequences $(\alpha_n) \in \{-1,+1\}^\infty$.

(3): Clearly any symmetric sequence (X_n) is even, and all its coordinate variables, X_n, are even. Conversely, if $X_1, X_2, \ldots$ are independent and even, then the sequence (X_n) is symmetric.

(4): Let E be a linear space and g a map from E into the extended positive line $\overline{R}_+ = [0,\infty]$. Then g is called <u>even</u> if $g(x) = g(-x)$, <u>subadditive</u> if $g(x+y) \leq g(x) + g(y)$, convex if $g(tx + (1-t)y) \leq t\, g(x) + (1-t)\, g(y)$ for all $0 \leq t \leq 1$, or <u>quasiconvex</u> if $g(\frac{1}{2}x + \frac{1}{2}y) \leq \max\{g(x), g(y)\}$. Note that any convex function is quasiconvex, and any seminorm is convex.

(5): Let E be a linear space, then E^* denotes its <u>algebraic dual</u>, i.e. the set of linear functionals from E into R. If $F \subseteq E^*$ then $(E, \sigma(F))$ is a measurable linear space and

(2.19) $$\sigma(F) = \mathcal{B}a(E, \sigma(E,F)) \subseteq \mathcal{B}(E, \sigma(E,F)).$$

But in general $\sigma(F)$ differs from the Borel σ-algebra of the $\sigma(E,F)$-topology.

(6): If $(E_t, \mathcal{B}_t)$ is a measurable linear space for every $t \in T$, then so is the product space $(E, \mathcal{B})$, where

$$E = \prod_{t \in T} E_t, \quad \mathcal{B} = \bigotimes_{t \in T} \mathcal{B}_t .$$

More generally, any projective limit of measurable linear spaces is a measurable linear space. To be precise: If $(E_t, \mathcal{B}_t)$ is a measurable linear space, E is a linear space, and p_t a linear map : $E \to E_t$ for all $t \in T$, then $(E, \mathcal{B})$ is a measurable linear space, if $\mathcal{B}$ is the smallest σ-algebra on E making p_t measurable for all $t \in T$.

(7): If (E,τ) is a separable pseudometrizable linear topological space, then $(E, \mathcal{B}(E))$ is a measurable linear space.

(8): If (E,τ) is an analytic linear topological space, then $(E, \mathcal{B}(E))$ is a measurable linear space.

(9): Let $(E,\mathcal{B})$ be a measurable linear space. If $\{0\} \in \mathcal{B}$ we say, that $(E, \mathcal{B})$ is <u>separated</u>. It is easily checked, that $(E,\mathcal{B})$ is a separated measurable linear space, if and only if there exists an injective measurable map $f: E \to [0,1]$. Thus a separated measurable linear space has cardinality equal to the continuum.

(10): Let $(E, \|\cdot\|)$ be a Banach space. If E is separable then $(E, \mathcal{B}(E))$ is a measurable linear space. By (9) above we see, that this need not to be the case, if we drop the separability condition. For instance Talagrand has shown, that $(\ell^\infty, \mathcal{B}(\ell^\infty))$ is a measurable linear space, if and only if the continuum hypothesis holds.

(11): If $(E, \mathcal{B})$ is a measurable linear space, and E_o is a linear subspace of E, then $(E_o, \mathcal{B}_o)$ is a measurable linear space, where $\mathcal{B}_o$ is <u>the trace</u> of $\mathcal{B}$ on E_o, i.e. $\mathcal{B}_o = \{B \cap E_o \mid B \in \mathcal{B}\}$.

IV. <u>Random vectors</u> (1): Let (E,τ) be a locally convex linear space. Then a map $X: \Omega \to E$ is called an E-<u>valued random vector</u>, if for every continuous seminorm $|\cdot|$ on E we have

(2.20) X is measurable: $(\Omega, F) \to (E, \mathcal{B}(E, |\cdot|))$, and the range $X(\Omega)$ is $|\cdot|$-separable

It is an easy exercise to show that, if S is a set of seminorms generating the topology on (E,τ), then X is an E-valued random vector, if and only if (2.20) holds for all seminorms $|\cdot|$ belonging to S.

(2): The reason for the separability condition in (2.20) can be found in III.(7) and III(10) above. This is the only way, we can assure, that a linear combination of random vetcors is a random vector.

(3): Let $(E, \|\cdot\|)$ be a Banach space and X an E-valued random vector. Then <u>the mean</u> or <u>the expectation</u> of X, denoted $E\,X$, is the Bochner integral (see [11, III.2])

$$E\,X = \int_\Omega X(\omega)\, P(d\omega) \tag{2.21}$$

provided of course that the integral exists. Note that the law of X is a Radon probability measure on E, and if $\mu = \mathcal{L}(X)$ then

$$E\,X = \int_E x\mu(d\,x)\,. \tag{2.22}$$

If μ is a Radon measure on E, then <u>the mean</u> of μ is defined to be the Bochner integral

$$\int_E x\,\mu(d\,x) \tag{2.23}$$

provided of course that it exists.

(4): The set of all <u>E-valued random vectors</u> is denoted $L_E^0(\Omega, F, P)$

or shortly $L^0_E(P)$. If $E = R$ we drop the subscript E, and write $L^0(\Omega, F, P)$ or $L^0(P)$. On $L^0_E(P)$ we define the metric : $\|X-Y\|_0$ by

(24) $$\|X\|_0 = E\,(\text{Arc tg}\,\|X\|) \qquad \forall X \in L^0_E(P).$$

(5): The numbers $E\|X\|^p$ are called <u>the moments</u> of $X(0 < p < \infty)$. The set of all <u>E-valued random vectors</u> with a <u>finite</u> <u>p-th moment</u> $(0 < p < \infty)$ is denoted $L^p_E(\Omega, F, P)$ or shortly $L^p_E(P)$. Again we drop the subscript E, when $E = R$. On $L^p_E(P)$ we define

(2.25) $$\|X\|_p = E\|X\|^p \quad \text{if } 0 < p \leq 1, \; X \in L^p_E(P)$$

(2.26) $$\|X\|_p = \{E\|X\|^p\}^{1/p} \quad \text{if } 1 \leq p < \infty, \quad X \in L^p_E(P)$$

Then $(L^p_E(P), \|\cdot\|_p)$ is a Fréchet space (not necessarily locally convex) for $0 \leq p < 1$, and $(L^p_E(P), \|\cdot\|_p)$ is a Banach space for $1 \leq p < \infty$.

(6): The set of all <u>essentially bounded</u>, <u>E-valued, random vectors</u> X (i.e. $\exists K > 0 : \|X\| \leq K$ a.s.) is denoted $L^\infty_E(\Omega, F, P)$ or shortly $L^\infty_E(P)$. If $E = R$ we drop the subscript E. $L^\infty_E(P)$ is a Banach space under the norm:

(2.27) $$\|X\|_\infty = \operatorname*{ess\,sup}_\omega \|X(\omega)\| = \inf\{a > 0 \mid \|X\| \leq a \text{ a.s.}\}$$

(7): Let $(E, \|\cdot\|)$ be a Banach space and X an E-valued random vector. Let E' be <u>the dual</u> of E, i.e. the set of all continuous linear functionals from E into R. If $\langle x', X\rangle \in L^2(P)$ for all $x' \in E'$ (e.g. if $X \in L^2_E(P)$), then we define <u>the covariance</u> r_X of X by the formula:

(2.28) $$r_X(x', y') = E\{\langle x', X\rangle \langle y', X\rangle\} \quad \forall x', y' \in E'.$$

It is easily checked, that r_X is <u>bilinear</u> (i.e. $r_X(\cdot, y')$ and $r_X(x', \cdot)$ are linear), <u>symmetric</u> (i.e. $r_X(x', y') = r_X(y', x')$), and <u>positive definite</u> (i.e. $r_X(x', x') \geq 0$). Such a function is called a <u>positive definite quadratic form</u>. If $\langle x', X\rangle X \in L^1_E(P)$ $\forall x' \in E'$, then we define <u>the covariance operator</u> R_X, of X by :

(2.29) $$R_X x' = E\{\langle x', X\rangle X\} \qquad \text{for } x' \in E'.$$

Then R_X is a continuous linear operator: $E' \to E$, such that

(2.30) $\quad r_X(x',y') = \langle x', R_X y'\rangle = \langle y', R_X x'\rangle \quad \forall x',y' \in E'$

(8): If μ is a Radon measure on E, then the <u>covariance</u> r_μ, and the <u>covariance operator</u> R_μ, are defined as follows:

(2.31) $\quad r_\mu(x',y') = \int_E \langle x',x\rangle \langle y',x\rangle \ \mu(dx)$

(2.32) $\quad R_\mu x' = \int_E \langle x',x\rangle x \ \mu(dx)$

provided of course that the relevant integrals exist. Note that if $X \sim \mu$ then $r_X = r_\mu$ and $R_X = R_\mu$.

(9): A map $g:E \to \overline{R}_+$ is said to be <u>measure convex</u>, if g is measurable and $g(E\,X) \leq E\,g(X)$ for all $X \in L^1_E(P)$. Clearly any measure convex function is convex. If $\dim E < \infty$, then every measurable convex function $g:E \to \overline{R}_+$ is measure convex (<u>Jensen's inequality</u>). If $\dim E = \infty$, this need not hold, but we have that every lowersemicontinuous convex function $g:E \to \overline{R}_+$ is measure convex.

(10): Let $X,Y \in L^1_E(P)$ be independent, let $m = E\,X$, and g a measure convex function: $E \to \overline{R}_+$. Then we have: $E\,g(m+Y) \leq E\,g(X+Y)$, by (2.9).

(11): A set $\mathcal{X}$ of E-valued random variables is said to be <u>stochastically bounded</u>, if $\mathcal{X}$ is a bounded subset of $L^0_E(P)$, i.e. if

(2.23) $\quad \forall \varepsilon > 0 \quad \exists \delta > 0 : \quad \|\delta X\|_0 \leq \varepsilon \quad \forall X \in \mathcal{X}$.

It is straightforward to verify that (2.23) is equivalent to any of the following three conditions:

(2.34) $\quad \forall \varepsilon > 0 \ \exists K < \infty : \quad P(\|X\| \geq K) \leq \varepsilon \quad \forall X \in \mathcal{X}$

(2.35) $\quad \exists \phi : R_+ \to R_+ \ \exists K < \infty$, so that ϕ is increasing,

$$\lim_{t\to\infty} \phi(t) = \infty \quad \text{and} \quad E\,\phi(\|X\|) \leq K \qquad \forall X \in \mathcal{X}$$

(2.36) $\quad \{\mathcal{L}(\|X\|) \mid X \in \mathcal{X}\}$ is a uniformly tight subset of $\mathrm{Prob}(R_+)$.

V <u>Characteristic functionals</u> (1): Let $(E, \|\cdot\|)$ be a Banach space and let E' be its dual. Then on E' we have the $\|\cdot\|$-<u>topology</u>:

(2.37) $$\|x'\| = \sup\{|<x', x>| : x \in E,\ \|x\| \leq 1\},$$

the w*-<u>topology</u> $\sigma(E', E)$, and <u>the topology of uniform convergence on compact subsets of</u> E, denoted $\Pi(E', E)$. I.e. $x' \to x'_o$ in $\sigma(E', E)$, if and only if

(2.38) $$\lim_{\alpha} <x'_\alpha, x> = <x'_o, x> \quad \forall x \in E .$$

And $x'_\alpha \to x'_o$ in $\Pi(E', E)$, if and only if

(2.40) $$\lim_{\alpha} \sup_{x \in K} |<x'_\alpha - x'_o, x>| = 0 \qquad \forall K \text{ compact} \subseteq E .$$

(2): It is well-known that the dual of E' under both $\sigma(E', E)$ and $\Pi(E', E)$ equals E. I.e. if $f: E' \to R$ is a linear functional, which is either $\sigma(E', E)$-continuous or $\Pi(E', E)$-continuous, then f is given by : $f(x') = <x', a> \ \forall x'$, for some $a \in E$.

(3): If X is an E-valued random vector, we define its <u>characteristic functional</u>, denoted ϕ_X, by

(2.41) $$\phi_X(x') = E\{e^{i<x', X>}\} \qquad \forall x' \in E' .$$

Similarly if μ is a finite Radon measure on E, we define its <u>Fourier transform</u> or in probabilistic terminology its <u>characteristic functional</u>, denoted $\hat{\mu}$, by

(2.42) $$\hat{\mu}(x') = \int_E e^{i<x, x'>} \mu(dx)$$

Note that $\phi_X = \hat{\mu}$ if $\mu = \mathcal{L}(X)$. If μ is a Radon probability measure on E, then we have

(2.43) $$\hat{\mu}(0) = 1 \quad , \quad |\hat{\mu}(x')| \leq 1 \qquad \forall x' \in E'$$

(2.44) $$\hat{\mu} \text{ is } \Pi(E', E)\text{-continuous on } E' .$$

(4): Let $\{\mu_\alpha\}$ be a net in Prob(E). Then clearly we have

(2.45) $$\mu_\alpha \overset{\sim}{\to} \mu \Longrightarrow \hat{\mu}(x') = \lim_\alpha \hat{\mu}_\alpha(x') \qquad \forall x' \in E'.$$

The converse implication (Levy's continuity theorem) is valid, whenever E is finite dimensional, but it fails for an infinite dimensional Hilbert space. However, let F be a linear subset of E, and take H to be the set of functions f of the form

(2.46) $$f(x) = \phi(\langle a_1', x\rangle, \ldots, \langle a_m', x\rangle)$$

for some $m \geq 1$, some $a_1', \ldots, a_m' \in F$ and some $\phi \in C(R^m)$. Then by (2.16) we have

(2.47) $$\hat{\mu}_\alpha(x') \to \hat{\mu}(x') \;\forall x' \in F \;\Rightarrow\; \mu_\alpha \overset{\sim}{\to} \mu \text{ in } (E, \sigma(E,F))$$

It would be tempting to take $H_o = \{e^{i\langle x', \cdot\rangle} \mid x' \in F\}$ above, but $\sigma(E, H_o)$ is not equal to $\sigma(E,F)$. Even when $E = R$ we have, that the functions: $t \to e^{ita}$ for $a \in R$, do not generate the usual topology on R. It is an amusing exercise to show, that there exists a net $\{n_\alpha\}$ of integers, such that $n_\alpha \to \infty$, but $\exp(ian_\alpha) \to 1$ for all $a \in R$.

VI Symmetrization (1): Let (E, B) be a measurable linear space, and X an E-valued random variable. Then a symmetrization of X is an E-valued random variable X^s, such that $X^s = X' - X''$ where X' and X'' are two independent copies of X. Clearly any symmetrization X^s of X is even.

(2): If $X_1, X_2, \ldots$ are independent and $X^s = (X_n^s)$ is a symmetrization of $X = (X_n)$. Then X^s is a symmetric sequence and $X_1^s, X_2^s, \ldots$ are independent.

(3): It will turn out, that the symmetrization is one of the main methods in what follows. The transition between a random variable X and a symmetrization X^s is given by the symmetrization inequalities:

(2.48) $$P(g(X) \leq a)\; E\,\phi[g(X)] \leq E\,\phi[g(X^s) + a]$$

(2.49) $$E\,\phi[g(X^s)] \leq E\,\phi[2\,g(X)] + E\,\phi[2\,g(-X)]$$

which holds for any $a \geq 0$, any subadditive measurable map $g: E \to \overline{R}_+$ and any increasing function ϕ from $\overline{R}_+$ into $\overline{R}_+$. Putting $\phi = 1_{]t,\infty]}$ we find

$$P(g(X) \leq a)\, P(g(X) > t) \leq P(g(X^s) > t-a) \tag{2.50}$$

$$P(g(X^s) > t) \leq P(g(X) > \tfrac{1}{2} t) + P(g(-X) > \tfrac{1}{2} t) \tag{2.51}$$

for all $a, t \geq 0$.

(4): Let $(E, \|\cdot\|)$ be a Banach space, X an E-valued random vector and g a measure convex map from E into $\overline{R}_+$. If X^s is a symmetrization of X, then we have (see IV.(10))

$$E\, g(X-m) \leq E\, g(X^s) \quad \text{where} \quad m = E\,X \tag{2.52}$$

$$E\, g(X^s) \leq \tfrac{1}{2}\, E\, g(2X) + \tfrac{1}{2}\, E\, g(-2X) \tag{2.53}$$

Putting $g(x) = \|x\|^p$ where $1 \leq p < \infty$, then g is measure convex and so

$$E\,\|X-m\|^p \leq E\,\|X^s\|^p \quad \text{where} \quad m = E\,X \quad \forall p \geq 1 \tag{2.54}$$

$$E\,\|X^s\|^p \leq 2^p\, E\,\|X\|^p \quad \forall p \geq 1 . \tag{2.55}$$

(5): The simplest possible non-trivial sequence of independent random variables is the <u>Bernoulli sequence</u> (ε_n), which is a sequence of independent identically distributed random variables $\varepsilon_1, \varepsilon_2, \ldots$ taking the values $+1$ and -1 with equal probabilities, i.e. $P(\varepsilon_n = +1) = P(\varepsilon_n = -1) = \frac{1}{2}$. The Bernoulli sequence is a model of an infinite series of coin tossing (head up:+1, tail up:-1). It is sometimes called the Rademacher sequence by functional analysists; however James Bernoulli studied this sequence 200 years before Rademacher was born, so I see no reason for changing the name.

(6): Despite of its simplicity the Bernoulli sequence reveals all the randomness of a symmetric sequence (X_n). The reason being, that $(X_n) \sim (\varepsilon_n X_n)$, whenever (X_n) is a symmetric sequence and (ε_n) a Bernoulli sequence independent of (X_n), and if so then

$$E\,\phi(X_1, X_2, \ldots) = \int_{E^\infty} E\,\phi(\varepsilon_1 x_1, \varepsilon_2 x_2, \ldots)\, \mu(dx_1, dx_2, \ldots) \tag{2.56}$$

where μ is the law of (X_n) on E^∞.

VII Stohastic processes (1): Let $X = \{X(\theta) \mid \theta \in \Theta\}$ be a real valued stochastic process. If $X(\theta) \in L^r(P)$ for all $\theta \in \Theta$, where $r \in [0,\infty]$, we say that X is an r-th order process. If X is a first order process we define its mean function: m_X, by

$$(2.57) \qquad m_X(\theta) = E\, X(\theta) \qquad \forall \theta \in \Theta$$

and we say that X is centered if $m_X \equiv 0$. If X is a second order process, we define its covariance function: σ_X, and its intrinsic metric : ρ_X, by

$$(2.58) \qquad \sigma_X(\theta,\tau) = E\, X(\theta)\, X(\tau) \qquad \forall \theta,\tau \in \Theta$$

$$(2.59) \qquad \rho_X(\theta,\tau) = \{E\,|X(\theta) - X(\tau)|^2\}^{1/2} \qquad \forall \theta,\tau \in \Theta .$$

Then σ_X is symmetric (i.e. $\sigma_X(\theta,\tau) = \sigma_X(\tau,\theta)$ τ,θ) and positive definite (i.e. the $n \times n$ matrix : $\{\sigma(\theta_p,\theta_q)\}$ is positive definite $\forall n \geq 1$ $\forall \theta_1,\ldots,\theta_n$), and ρ_X is a pseudometric on Θ satisfying:

$$(2.60) \qquad \rho_X(\theta,\tau) = \{\sigma_X(\theta,\theta) + \sigma_X(\tau,\tau) - 2\,\sigma_X(\theta,\tau)\}^{1/2} \qquad \forall \theta,\tau .$$

(2): A stohastic process $X = \{X(\theta) \mid \theta \in \Theta\}$ is said to be gaussian, if all linear combinations $\Sigma\, t_j\, X(\theta_j)$ have 1-dimensional gaussian law (we allow a gaussian law to have 0 variance, thus any Dirac measure is a gaussian law). Hence, if X is gaussian, then X is an r-th order process for all $0 < r < \infty$, and

$$(2.61) \qquad E\exp\Big[\sum_{k=1}^{n} c_k\, X(\theta_k)\Big] = \exp\Big[\sum_{k=1}^{n} c_k m(\theta_k) + \frac{1}{2}\sum_{j=1}^{n}\sum_{k=1}^{n} c_j c_k \sigma(\theta_j,\theta_k)\Big]$$

for all complex numbers $c_1,\ldots,c_n$ and all $\theta_1,\ldots,\theta_n \in \Theta$, where m is the mean function and σ the covariance function of X.

(3): If $(E,\|\cdot\|)$ is a Banach space and X is an E-valued random vector, then $X(x') = \langle x',X\rangle$ for $x' \in E'$ is a real valued stohastic process with time set E'. And, in complete analogy with the definition above we shall say that X is gaussian if $\langle x',X\rangle$ has a 1-dimensional gaussian law for all $x' \in E'$. Now suppose that X is a centered (i.e. $E\, X = 0$) gaussian E-valued random vector, then its characteristic functional ϕ is given by (cf. (2.61))

(2.62) $$\phi(x') = E\, e^{i<x',X>} = e^{-\frac{1}{2}r(x',x')} \quad \forall x' \in E'$$

where r is the covariance of X .

(4): There are several equivalent ways of defining centered gaussian random vectors. Let X be an E-valued random vector and Y an independent copy of X , then it can be shown that the following five statements are equivalent

(2.63) $<x',X>$ has a centered gaussian law $\forall x' \in E'$

(2.64) $T X$ has a centered gaussian law (in R^m) for all continuous linear operators $T:E \to R^m$ $\forall m \geq 1$

(2.65) $\phi_X(x') = e^{-\frac{1}{2} r(x',x')}$ for some positive definite quadratic form r on $E' \times E'$

(2.66) $X + Y \sim \sqrt{2}\, X$

(2.67) $X + Y$ and $X - Y$ are independent and indentically distributed.

Note that (2.63)-(2.65) may be used to define a centered gaussian random vector in a locally convex space, but if E is not locally convex, then E' may degenerate to $\{0\}$, and (2.63)-(2.65) cannot be used. However (2.66) may be used to define a centered gaussian variable with values in a linear topological space, and (2.67) may even be used to define a centered gaussian variable with values in a topological group.

(5): Condition (2.65) may be used to define a centered gaussian measure on a Banach space $(E,\|\cdot\|)$. A Radon measure μ on E , is said to be a centered gaussian measure on E if there exists a positive definite quadratic form r on $E' \times E'$, such that

(2.68) $$\hat{\mu}(x') = e^{-\frac{1}{2} r(x',x')} \quad \forall x' \in E' .$$

It can be shown, that if (2.68) holds, then μ has mean 0 , and r is the covariance of μ. This definition naturally rasies the question: "Which positive definite quadratic form or $E' \times E'$ are gaussian ,i.e.

satisfy (2.68) for some Radon measure μ on E?" The answer to this question is not known in general, and the study of gaussian quadratic forms is an important and difficult subject. In a Hilbert space the answer is : "$r(x,y) = \langle Ax, Ay \rangle$ for some positive definite Hilbert-Schmidt operator A".

(6): A real valued stochastic process $X = \{X(\theta) | \theta \in \Theta\}$ may be considered as an R^Θ-valued random variable, where R^Θ is the product of Θ copies of the real line, with its product σ-algebra $\mathcal{B}^\Theta$. Note that $(R^\Theta, \mathcal{B}^\Theta)$ is a measurable linear space by III.(6). Thus a <u>version</u> of X is a stochastic process Y such that the law of Y (on R^Θ) equals the law of X , i.e. such that $(X(\theta_1),\ldots,X(\theta_n)) \sim (Y(\theta_1),\ldots,Y(\theta_n))$ for all $n \geq 1$ and all $\theta_1,\ldots,\theta_n \in \Theta$.

(7): My main motivation for studying probability in Banach spaces, was the possibility of applying the general theory to stochastic processes.

Let E_o be a linear subset of R^Θ (=the set of all real-valued functions on Θ) and let $\|\cdot\|$ be a norm on E_o . Then the completion $(E,\|\cdot\|)$ of $(E_o,\|\cdot\|)$ is a Banach space, and if $X = \{X(\theta)|\theta \in \Theta\}$ is a stochastic process satisfying

(2.69) $$X(\omega,\cdot) \in E_o \quad \forall\omega$$

(2.70) $$\{X(\omega,\cdot) | \omega \in \Omega\} \quad \text{is } \|\cdot\|\text{-separable}$$

(2.71) $$\omega \to X(\omega,\cdot) \quad \text{is measurable: } (\Omega,F) \to (E, \mathcal{B}(E))$$

then X is an E-valued random vector. A process satisfying (2.69)-(2.71) is called an <u>E_o-valued process</u>. The conditions (2.69) and (2.71) are fairly innocent, but the separability condition (2.70) may be a quite severe restriction.

(8): Let E_o be a linear subset of R^Θ with a norm $\|\cdot\|$, and let $(E,\|\cdot\|)$ be the completion of $(E_o,\|\cdot\|)$. We shall then normally require, that the integral of an E_o-valued process can be computed pointwise . More precisely, a subset E_1 of E_o , is said to be <u>functionally closed</u> in E if every E_o-valued process $X \in L^1_E(P)$, satisfying $P(X \in E_1)=1$ has the property

(2.72) $$EX = m_X \in E_1$$

where m_X is the mean function of X, and EX is the mean of X as an E-valued random vector.

(9): Let $(E_o, \|\cdot\|)$ be a normed linear subspace of R^Θ, and let $(E, \|\cdot\|)$ be its completion. If E_1 is a subset of E_o satisfying

(2.73) $\quad E_1$ is a $\|\cdot\|$-closed convex subset of E, and $f \to f(\theta_o)$ is $\|\cdot\|$-continuous : $E_1 \to R$ for all $\theta_o \in \Theta$

then E_1 is functionally closed in E. This shows that the spaces $(B(\Theta), \|\cdot\|_\infty)$, $(C(\Theta), \|\cdot\|_\infty)$ and $(Lip(\Theta,\rho), \|\cdot\|_\rho)$ are all functionally closed. But also the spaces like $(L^p(\Theta, A, \nu), \|\cdot\|_p)$ for $1 \leq p \leq \infty$ are functionally closed even though they do not satisfy (2.73).

If (T, ρ) is a pseudometric space and $\Theta = \mathcal{B}(T)$, then $E_o = ca_\tau(T)$ is a linear subset of R^Θ, and $\|\cdot\|_\rho$ is a norm on E_o. Here E is the $\|\cdot\|_\rho$-closure of E_o in $M(T,\rho)$, and E is not necessarily a subset of R^Θ. In this case it is easily checked that $E_1 = ca_\tau^+(S)$ is functionally closed in E, but again (2.73) is not necessarily satisfied.

(10): Let $(E_o, \|\cdot\|)$ be a normed linear subset of R^Θ, with completion $(E, \|\cdot\|)$, and suppose that E_o is functionally closed in E. Let X be an E-valued second order E_o-valued process with $E\|X\|^2 < \infty$. Then it is easily checked that the covariance operator R of X, is given in terms of the covariance function σ of X by the formula

$$(2.74) \qquad (Rx')(\theta) = \langle x', \sigma(\theta, \cdot) \rangle \qquad x' \in E' \quad \forall \theta .$$

I.e. $\sigma_\theta = \sigma(\theta, \cdot)$ belongs to E_o $\forall\theta$, and $\phi(\theta) = \langle x', \sigma_\theta \rangle$ belongs to E_o, and $Rx' = \phi$. Thus, we find that R maps E' into E_o.

(11): Let $(E, \|\cdot\|)$ be a normed linear subset of R^Θ, and σ a symmetric positive definite map: $\Theta\times\Theta \to R$. Then we say that σ is <u>E-gaussian</u>, if there exists a centered E-valued gaussian process with covariance σ. In particular if $E = B(\Theta)$ or $E = C(\Theta)$ (with respect to some topology on Θ), we use the terminology B-<u>gaussian</u> or C-<u>gaussian</u>. It is one of the very difficult problems of probability theory to give good necessary and / or sufficient conditions for a symmetric positive definite function σ to be B-gaussian or C-gaussian.

(12): Similarly, we say that a pseudometric ρ on Θ is E-gaussian , if there exists a centered E-valued gaussian process with intrinsic metric ρ . And we use the terminology B-gaussian or C-gaussian pseudometric as above, i.e. when $E = B(\Theta)$ or $E = C(\Theta)$.

3. MAXIMAL INEQUALITIES

We shall in this section derive some important maximal inequalities. A maximal inequality is an inequality stating that the maximum, $M = \max_{1\le j\le n} S_j$, of n random variables $S_1,\dots,S_n$ is probabilisticly speaking of the same order of magnitude as the last random variable S_n.

Let us fix the setting of this section: $(E_n, \mathcal{B}_n)$ is a measurable linear space for $n = 1,2,\dots,$

$$E^\infty = \prod_{j=1}^{\infty} E_j \quad , \qquad \mathcal{B}^\infty = \bigotimes_{j=1}^{\infty} \mathcal{B}_j$$

and q is a measurable map: $E^\infty \to \overline{R}_+$. We shall then consider the random variables

$$(3.1) \qquad S_n = q(X_1,X_2,\dots,X_n,0,0\ \dots)$$

$$(3.2) \qquad M_n = \max\{S_1,\dots,S_n\} \ , \ M = \sup_j S_j \ .$$

And we shall compare M_n with S_n.

THEOREM 3.1. <u>Let</u> (X_n) <u>be a symmetric</u> E^∞<u>-valued random sequence, and let</u> S_n, M_n <u>and</u> M <u>be defined as in (3.1) and (3.2). If</u> q <u>is quasiconvex, and</u> $\phi:\overline{R}_+\to\overline{R}_+$ <u>is increasing, then we have</u>

$$(3.1.1) \qquad E\,\phi(S_n) \le E\,\phi(M_n) \le 2\,E\,\phi(S_n)$$

$$(3.1.2) \qquad \sup_n E\,\phi^-(S_n) \le E\,\phi^-(M) \le 2\,\liminf_n E\,\phi(S_n)$$

$$(3.1.3) \qquad M<\infty \text{ a.s.} \Longleftrightarrow \{S_n\} \ \underline{\text{is stochastically bounded}}$$

<u>where</u> $\phi^-(x) = \lim_{y\uparrow x} \phi(y) = \sup_{y<x} \phi(y)$.

<u>Moreover if</u> $S_n \overset{\sim}{\to} S$, <u>then we have</u>

$$(3.1.4) \qquad E\,\phi(S) \le E\,\phi(M) \le 2\,E\,\phi(S)$$

PROOF (3.1.1): A standard "integration by parts"-argument shows that if suffices to prove the following

(i) $\qquad P(M_n > t) \leq 2\, P(S_n > t) \qquad \forall t \geq 0 .$

So let $t \geq 0$ be given, and consider the first entrance of $\{S_n\}$ into the interval $]t, \infty]$:

$$\tau = \inf \{ j \geq 1 \mid S_j > t \} \qquad (\inf \emptyset = \infty)$$

If $1 \leq j \leq n$ we define

$$Z_{nj} = (X_1, \ldots, X_j, -X_{j+1}, \ldots, -X_n, 0, 0, \ldots) .$$

Then we have $Z_{jj} = \frac{1}{2} Z_{nj} + \frac{1}{2} Z_{nn}$ and $S_n = q(Z_{nn})$. So by the quasi-convexity of q we find:

$$S_j = q(\tfrac{1}{2} Z_{nj} + \tfrac{1}{2} Z_{nn}) \leq \max \{ q(Z_{nj}), S_n \} .$$

Now if $\tau = j$ then $S_j > t$, so either $S_n > t$ or $q(Z_{nj}) > t$, and thus we conclude

(ii) $\qquad P(\tau = j) \leq P(\tau = j,\ S_n > t) + P(\tau = j, q(Z_{nj}) > t).$

Now note that

$$\{\tau = j\ , S_n > t\} = \{ Z_{nn} \in A_{jn} \}$$

$$\{\tau = j\ , q(Z_{nj}) > t\} = \{ Z_{nj} \in A_{jn} \}$$

where

$$A_{jn} = \left\{ (x_j) \in E^\infty \;\middle|\; \begin{array}{l} q(x_1, \ldots, x_i, 0, \ldots) \leq t \qquad 1 \leq i \leq j-1 \\ q(x_1, \ldots, x_j, 0, \ldots) > t\ ,\ q(x_1, \ldots, x_n, 0, \ldots) > t \end{array} \right\}$$

By the symmetry of (X_n) we have that $Z_{nn} \sim Z_{nj}$, so the two probabilities on the left hand side of (ii)are equal. Thus we find

$$P(\tau \leq n) = \sum_{j=1}^{n} P(\tau = j) \leq 2 \sum_{j=1}^{n} P(\tau = j,\ S_n > t)$$

$$\leq 2\, P(S_n > t)$$

and since $\{\tau \leq n\} = \{M_n > t\}$ we see that (i) and thereby (3.1.1) is proved.

(3.1.2) follows from (3.1.1) by letting $n \to \infty$.

(3.1.3) follows evidently from (3.1.2).

The proof of (3.1.4) is simple, and the details are left to the reader. □

Notice that Theorem 3.1 states, roughly speaking, that the maximum of $S_1, \ldots, S_n$ is at most $2\,S_n$ (not a.s. but in law). Now using the symmetrization inequalities (2.48)-(2.51) we obtain

THEOREM 3.2. Let $\{X_n\}$ be independent E_n-valued random variables, and let S_n, M_n and M be defined as in (3.1) and (3.2). If q is even, subadditive and quasiconvex, and $\phi : \overline{R}_+ \to \overline{R}_+$ is increasing, then for all $a > 0$ we have

(3.2.1) $$P(M_n \leq a)\ E\,\phi(M_n) \leq 4\ E\,\phi(S_n + a)$$

(3.2.2) $$P(M \leq a)\ E\,\phi^-(M) \leq 4\ \liminf_n E\,\phi(S_n + a)$$

(3.2.3) $$M < \infty \text{ a.s.} \iff \{S_n\} \text{ is stochastically bounded}$$

where $\phi^-(x) = \lim_{y \uparrow x} \phi(y) = \sup_{y < x} \phi(y)$.

Moreover if $S_n \overset{\sim}{\to} S$, then we have

(3.2.4) $$P(M \leq a)\ E\,\phi(M) \leq 4\ E\,\phi(S + a)$$

(3.2.5) $$E\,\phi(S) \leq E\,\phi(M)$$

PROOF. The proof is a routine application of the symmetrization inequalities (2.48)-(2.51), and I shall leave the details to the reader. □

In applications of Theorem 3.2 one usually take a to be a median of M, i.e. a real number satisfying

$$P(M < a) \leq \tfrac{1}{2} \leq P(M \leq a)$$

(If $M < \infty$ a.s. then M admits at least one median). If a is a median of M, then (3.2.1), (3.2.2) and (3.2.4) take the form:

$$(3.2.1^*) \qquad E\,\phi(M_n) \leq 8\,E\phi(S_n+a)$$

$$(3.2.2^*) \qquad E\,\phi^{-}(M) \leq 8 \liminf_n E\,\phi(S_n+a)$$

$$(3.2.4^*) \qquad E\,\phi(M) \leq 8\,E\,\phi(S+a)$$

So in this case, M_n is in law at most $8(S_n+a)$, and M is in law at most $8(S+a)$.

We shall derive a third and more striking maximal inequality. Put

$$(3.3) \qquad Y_n = q(0,\dots,0,X_n,0,\dots)\ ,$$

$$(3.4) \qquad N_n = \max\{Y_1,\dots,Y_n\} \text{ and } N = \sup_n Y_n\ .$$

If q is subadditive and even, then $Y_n \leq S_n + S_{n-1}$ and so $Y_n \leq 2M_n$. We shall now show, that M_n is dominated in law by $2N_n$ plus an exponentially decreasing term:

THEOREM 3.3. <u>Let $\{X_n\}$ be independent symmetric E_n-valued random sequence, and let S_n, M_n, M, Y_n, N_n and N be defined as in (3.1)-(3.4). If q is subadditive and quasiconvex, then for every $s,t,u \geq 0$ we have</u>

$$(3.3.1) \qquad P(S_n > s+t+u) \leq P(N_n > s) + 2P(S_n > t)\,P(S_n > u)$$

$$(3.3.2) \qquad P(M > s+t+u) \leq 2P(N > s) + 4P(M > t)\,P(M > u).$$

PROOF (3.3.1): As before we consider the first entrance of $\{S_j\}$ to $]t,\infty]$:

$$\tau = \inf\{j \geq 1 \mid S_j > t\} \qquad (\inf \emptyset = \infty).$$

Then $S_n > s+t+u$ implies $\tau \leq n$, so we have

$$(i) \qquad P(S_n > s+t+u) = \sum_{j=1}^{n} P(\tau = j,\ S_n > s+t+u).$$

Let $X_{jn} = (0,\dots,0,X_j,\dots,X_n,0,\dots)$, then

$$X_{1n} = X_{1j-1} + X_{jj} + X_{j+1n}$$

so by the suadditivity of q we have (note that $S_n = q(X_{1n})$ and $Y_n = q(X_{nn})$) :

$$S_n \leq S_{j-1} + Y_j + q(X_{j+1n}) \leq S_{j-1} + N_n + q(X_{j+1n}).$$

Hence if $\tau = j$, $N_n \leq s$ and $S_n > s+t+u$, then $S_{j-1} \leq t$ and $q(X_{j+1n}) > u$. Then we find

(ii) $$P(\tau = j,\ S_n > s+t+u) \leq P(\tau = j, N_n > s) + P(\tau = j,\ Y_{jn} > u)$$

where $Y_{jn} = q(X_{j+1n})$. Now Y_{jn} is a function of $(X_{j+1},\dots,X_n)$ and $\{\tau = j\}$ is a function of $(X_1,\dots,X_j)$, so $\{\tau = j\}$ and Y_{jn} are independent. Thus by Theorem 3.1 with $\phi = 1_{]u,\infty]}$, $\tilde{q}(x) = q(x_n,\dots,x_1,x_{n+1},\dots)$ and $\tilde{X} = (X_n,\dots,X_1,X_{n+1},\dots)$ we obtain

$$P(\tau = j,\ Y_{jn} > u) = P(\tau = j)\ P(Y_{jn} > u) \leq 2P(\tau = j)\ P(S_n > u).$$

Inserting this into (ii) and summing up over $j = 1,\dots,n$, we obtain (3.3.1).

(3.3.2) follows easily from (3.3.1) and the maximal inequality (3.1.2). □

Again a routine application of the symmetrization inequalities (2.48)-(2.51) allows us to drop the symmetry condition on X_n. We leave the details of the proof to the reader.

THEOREM 3.4. Let $\{X_n\}$ be independent E_n-valued random variables, and let S_n, M_n, M, Y_n, N_n and N be defined as in (3.1)-(3.4). If q is even, subadditive and quasiconvex, then there exists a decreasing function $F: R_+ \to [0,1]$ such that

(3.4.1) $$P(M \leq a)P(M > t+a) \leq F(t) \leq 2\ P(M > \tfrac{1}{2}\ t)$$

(3.4.2) $$F(2s+t+u) \leq 4P(N > s) + 4F(t)\ F(u)$$

for all $a,s,t,u \in R_+$. Thus, if a is a median for M , we have

(3.4.3) $$\tfrac{1}{2} P(M > t+a) \leq F(t) \leq 2P(M > \tfrac{1}{2}\ t) \qquad \forall t \geq 0.$$

Actually if (X_n^s) is a symmetrization of (X_n) we can take F to be the function:

(3.4.4) $$F(t) = P(\sup_n q(X_1^s,\dots,X_n^s,0,\dots) > t) \quad \forall t \geq 0 .$$

Let us consider a typical example, to which the maximal inequalities apply:

THEOREM 3.5. *Let* $(E, \mathcal{B})$ *be a measurable linear space,* $\{X_n\}$ *a sequence of independent* E-*valued random variables, and* $\{q_n\}$ *a sequence of measurable, even, subadditive, quasiconvex functions from* E *into* $\overline{R}_+$, *such that* $q_1 \leq q_2 \leq \dots$. *Let*

(3.5.1) $$q(x) = \sup_n q_n(x)$$

(3.5.2) $$S_n = X_1 + \dots + X_n \quad \text{and} \quad M = \sup_n q(S_n) .$$

If S *is an* E-*valued random variable, such that*

(3.5.3) $$q_i(S_n) \overset{\sim}{\to} q_i(S) \quad \text{as} \quad n \to \infty \ \forall i \geq 1 .$$

Then for every increasing map $\phi : \overline{R}_+ \to \overline{R}_+$ *we have*

(3.5.4) $$P(M \leq a) \cdot E\,\phi(M) \leq 4E\,\phi(q(S)+a) \qquad \forall a \geq 0$$

(3.5.5) $$E\,\phi(q(S)) \leq E\,\phi(M) .$$

If moreover X_n *is even for all* $n \geq 1$, *then we have*

(3.5.6) $$E\,\phi(q(S)) \leq E\,\phi(M) \leq 2E\,\phi(q(S)).$$

PROOF. Put $S_n^i = q_i(S_n)$, $M^i = \sup_n q_i(S_n)$ and $S^i = q_i(S)$. If $x = (x_j)$ we put

$$r_i(x) = \begin{cases} q_i(\Sigma x_j) & \text{if } \{j \mid x_j \neq 0\} \text{ is finite} \\ \infty & \text{otherwise} . \end{cases}$$

Then r_i is measurable, even, subadditive and quasiconvex on E^∞, and

$$r_i(X_1,\dots,X_n,0,\dots) = S_n^i , \qquad M^i = \sup_n S_n^i .$$

Thus by Theorem 3.2. we have

$$P(M^i \leq a) \cdot P(M^i > t) \leq 4P(S^i > t-a)$$

$$\leq 4P(q(S) > t-a)$$

since $S^i \leq q(S)$. Now $M^i \uparrow M$, so letting $i \to \infty$ we obtain

$$P(M \leq a)\ P(M > t) \leq 4P(q(S\ )+a > t)\ ,$$

from where (3.5.4) follows immediately.

(3.5.5) and (3.5.6) follow in a similar manner from (3.2.5) and (3.1.4).□

Let me make a few comments on the inequalities (3.3.2) and (3.4.2). Let $G(s) = P(N > s)$ be the "tail" distribution of N. Let $\phi : R_+ \to R_+$ be increasing with $\phi(0) = 0$, then by integration by parts we obtain

(3.5) $$E\ \phi(N) = \int_0^\infty G(s)\ d\,\phi\,(s)\,.$$

So the rate of decrease of G determines, whether $E\,\phi(N)$ is finite or infinite. If $X_1, X_2, \ldots$ are independent, then so are $Y_1, Y_2, \ldots$, and a little computation involving the Borel Cantelli lemmas shows that $E\,\phi\,(N) < \infty$, if and only if

(3.6) $$\sum_{n=1}^\infty \int_{\{Y_n > a\}} \phi(Y_n)\ dP < \infty$$

for some $a > 0$ (or, equivalently, for all $a > 0$). Since $\phi\,(Y_n)$ is a function of X_n, this means that knowing the law of X_n, we have a simple method of checking $E\,\phi\,(N) < \infty$.

Now suppose $M < \infty$ a.s., and let F be the function from Theorem 3.4. If ϕ is increasing, then by (3.4.3) we have

(3.7) $$E\,\phi\,(2M) < \infty \Rightarrow \int_0^\infty F(t)\ d\,\phi\,(t) < \infty \Rightarrow E\,\phi\,(M-a) < \infty\,.$$

So if $\phi(2t) \leq K\,\phi\,(t)$, then

(3.8) $$E\,\phi\,(M) < \infty \Leftrightarrow \int_0^\infty F(t)\ d\,\phi\,(t) < \infty$$

So again the rate of decrease of F determines the finiteness of

$E\,\phi(M)$. By (3.4.2) we have

$$(3.9)\qquad F(2s+t+u) \leq 4G(s)+4F(t)\,F(u) \qquad \forall\, s,t,u \geq 0 .$$

If $G(s) = 0$ for $s \geq s_o$ (i.e. $N \leq s_o$ a.s.) then by (3.9) we have

$$(3.10)\qquad F(t) \leq \beta(u)\, e^{-t\,\alpha(u)} \quad \forall u \geq 0 \quad \forall t \geq 2s_o+u \quad \text{where}$$

$$\beta(u) = 1/\,(4F(u)) \quad \text{and} \quad \alpha(u) = -(\log 4F(u))/(2s_o+u).$$

Thus, in general we may expect, that F is dominated by $4G$ plus exponentially fast decreasing term. With a bit of work one can actually show, that if ϕ is increasing and $\phi(2t) \leq K\,\phi(t)$, then

$$(3.11)\qquad E\,\phi(M)<\infty \Longleftrightarrow E\,\phi(N)<\infty \Longleftrightarrow \sup_n E\,\phi(S_n)<\infty .$$

If moreover $S_n \overset{\sim}{\to} S$, then we may add the following

$$(3.12)\qquad E\,\phi(M)<\infty \Longleftrightarrow E\,\phi(S)<\infty .$$

Since $\phi(t) = t^p$ satisfies $\phi(2t) = 2^p t^p$, this takes care of the moments of S,M and N . However the exponential functions $\phi(t) = \exp(\alpha t^p)$, do <u>not</u> satisfy this condition. But with somewhat more care, it can be shown that if

$$(3.13)\qquad E\exp(\alpha N^p)<\infty \quad \text{for some } \alpha, p>0$$

then there exists $\beta > 0$ such that

$$(3.14)\qquad \begin{cases} E\exp(\beta M^p) < \infty & \text{if } 0<p<1 \\ E\exp\left(\dfrac{\beta M}{\log^+ M}\right) < \infty & \text{if } p=1 \\ E\exp(\beta M) < \infty & \text{if } p>1 . \end{cases}$$

So, in this way we may also obtain exponential inequalities from (3.3.2) and (3.4.2).

4. SUMS OF INDEPENDENT RANDOM VARIABLES

In this section we shall study the convergence and boundedness of the partial sums:

$$S_n = X_1 + \dots + X_n$$

of a sequence of independent Banach space valued random vectors $\{X_1, X_2, \dots\}$. The main feature of such partial sums is that even very weak conditions on the boundedness or convergence of $\{S_n\}$ will imply very strong forms of convergence of $\{S_n\}$.

THEOREM 4.1. *Let* $\{X_n\}$ *be independent* E*-valued random vectors and* μ *a Radon probability measure on* E *, such that*

(4.1.1) $$\lim_{n\to\infty} E\, e^{i\langle x', S_n\rangle} = \hat{\mu}(x') \qquad \forall x' \in F$$

where $S_n = X_1 + \dots + X_n$ *and* F *is a linear subspace of* E' *separating points of* E (*i.e.* $\forall x \neq 0\ \exists x' \in F$ *such that* $\langle x', x\rangle \neq 0$). *Then there exists a sequence* $\{m_n\} \subseteq E$ *such that*

(4.1.2) $$\sum_{n=1}^{\infty} (X_n - m_n) \quad \text{converges in } \|\cdot\| \text{ a.s.}$$

(4.1.3) $$m = \sum_{n=1}^{\infty} m_n \quad \text{converges in the } \sigma(E,F)\text{-topology}.$$

And if $\phi : R_+ \to R_+$ *increases* , $\phi(0) = 0 = \lim_{t\to 0} \phi(t)$, *and* $\phi(2t) \leq K\phi(t)$, *then the following six statements are mutually equivalent:*

(4.1.4) $$E\{\sup_n \phi(\|X_n - m_n\|)\} < \infty$$

(4.1.5) $$E\{\sup_n \phi(\|\sum_{j=1}^{n} (X_j - m_j)\|)\} < \infty$$

(4.1.6) $$\sup_n E\, \phi(\|\sum_{j=1}^{n} (X_j - m_j)\|) < \infty$$

(4.1.7) $$E\, \phi(\|\sum_{j=1}^{\infty} (X_j - m_j)\|) < \infty$$

(4.1.8) $$\int_E \phi(\|x\|)\, \mu(dx) < \infty$$

(4.1.9) $$E_\phi\left(\left\|\sum_{j=n}^{\infty}(X_j-m_j)\right\|\right)\xrightarrow[n\to\infty]{} 0 \text{ and is finite } \forall n\geq 1 .$$

PROOF. The proof is going to be fairly long so let me introduce some preliminary notation and remarks.

First of all since X_n is separably valued for every n, and μ is a Radon measure, without loss of generality we may assume that E is separable. There exists a compact convex symmetric set $K\subseteq E$, such that

(i) $$\mu(E_K) = 1 \text{ , where } E_K=\bigcup_{n=1}^{\infty} nK =\{ x \mid \|x\|_K < \infty \}$$

and $\|\cdot\|_K$ is the Minkowski functional of K :

$$\|x\|_K = \inf\{ \lambda \geq 0 \mid x \in \lambda K\}.$$

Since E is separable, then so is every subset of E' in the w*-topology $\sigma(E',E)$. Let K^o be the polar of K in F, i.e.

$$K^o =\{ x' \in F : |\langle x',x\rangle| \leq 1 \ \forall x\in K\}$$

and let $\{ c_j' \mid j\geq 1\}$ be a countable $\sigma(E',E)$-dense subset of K^o. We put

$$q_i(x) = \max_{1\leq j\leq i} |\langle c_j' ,x\rangle| \qquad \forall x\in E .$$

Then we have

(ii) $\{c_j' \mid j\geq 1\}$ separates points of E

(iii) $q_1 \leq q_2 \leq \dots$, $q_i \uparrow \|\cdot\|_K$.

To see this let $x_o\in E$ and let $\|x_o\|_K > a\geq 0$, then $x_o\notin aK$ and aK is $\|\cdot\|$-compact and thus $\sigma(E,F)$-closed (F separates points, so $\sigma(E,F)$ is Hausdorff). By the Hahn-Banach theorem, there exists $x_o'\in F$ with $|\langle x_o',x_o\rangle| > a$ and $|\langle x_o',x\rangle|\leq a$ for all $x\in aK$. But then $x_o' \in K^o\cap F$, and so for some $i\geq 1$ we have $q_i(x \ \geq|\langle c_i',x_o\rangle| > a$. This shows that $\|\cdot\|_K\leq \sup q_i$, and the converse inequality is evident. Finally, note that (iii) implies (ii).

Now let us introduce some more notation. Let $(Y_n) = (X'_n - X''_n)$ be a symmetrization of (X_n), and let ν be the symmetrization of μ, i.e.

$$\nu(A) = \int_E \mu(x-A)\ \mu(dx) = \int_E \mu(dx) \int_E 1_A(x-y)\ \mu(dy).$$

Put

$$T_n = \sum_1^n Y_j = \sum_1^n (X'_j - X''_j)$$

and let T be an E-valued random vector with $\mathcal{L}(T) = \nu$.

Then we have (see(i) and (4.1.1)).

(iv) $$\nu(E_K) = \int_{E_K} \mu(x-E_K)\ \mu(dx) = \mu(E_K)^2 = 1$$

(v) $$\hat{\nu}(x') = |\hat{\mu}(x')|^2$$

(vi) $$E\{e^{i\langle x', T_n\rangle}\} = |E\{e^{i\langle x', S_n\rangle}\}|^2 \xrightarrow[n\to\infty]{} \hat{\nu}(x').$$

Thus, by (2.47) we conclude that

(vii) $T_n \overset{\sim}{\to} \nu$ when E is equipped with the $\sigma(E,F)$-topology.

Now, applying Theorem 3.5 (in particular(3.5.6)) to the sequences $\{Y_n\}$ and $\{q_n\}$ we obtain

(viii) $$P(L > t) \leq 2\nu(x \mid \|x\|_K > t)\ \forall t, \text{ where } L = \sup_n \|T_n\|_K.$$

By (iv) we have, that $\nu(x \mid \|x\|_K = \infty) = 0$, so (viii) shows, that $L < \infty$ a.s. Since $T_n(\omega) \in L(\omega)K$ for all n and ω, we have

(ix) $\{T_n(\omega) \mid n \geq 1\}$ is relatively compact a.s.

By the real valued equivalence theorem (see e.g.[18, Theorem B p.251]) we have that

(x) $$\lim \langle x', T_n\rangle = t(x') \quad \text{exists a.s.} \quad \forall x' \in F$$

(xi) $$\lim \langle x', S_n\rangle = s(x') \quad \text{exists a.s.} \quad \forall x' \in F.$$

Combining (ii), (ix) and (x) it follows, that T_n converges a.s. By the Fubini-Tonelli's theorem, it follows that

$$1 = P\left(\sum_{1}^{\infty} (X_j' - X_j'') \text{ converges}\right)$$

$$= \int_{E^\infty} P\left(\sum_{1}^{\infty} (X_j - y_j) \text{ converges}\right) \xi(dy_1, dy_2, \ldots)$$

where ξ is the law of (X_j). Now, note that the integrand is ≤ 1 and ξ is a probability measure, so the integrand must be 1 ξ- a.s. Thus, there exists $\{m_j\} \subseteq E$, such that

(xii) $\quad R = \sum_{1}^{\infty} (X_j - m_j) \quad$ converges a.s.

Now put

$$R_n = \sum_{1}^{n} (X_j - m_j), \qquad s_n = \sum_{1}^{n} m_j, \quad \lambda = \mathcal{L}(R).$$

Then $s_n = S_n - R_n$, so by (xi) and (xii) we have that

(xiii) $\quad f(x') = \lim_{n\to\infty} \langle x', s_n \rangle$ exists for all $x' \in F$,

(xiv) $\quad f$ is a linear functional : $F \to R$

(xv) $\quad \hat{\mu}(x') = e^{if(x')} \hat{\lambda}(x') \qquad \forall x' \in F$.

Let $0 < \varepsilon < \frac{1}{2}$ be given. Then by (2.4 4) there exists a $\|\cdot\|$-compact set $C \subseteq E$, such that $|\hat{\mu}(x') - 1| \leq \varepsilon$ and $|\hat{\lambda}(x') - 1| \leq \varepsilon$ for all $x' \in C^o$ (the polar of C in E'). So if $x' \in F \cap C^o$ then by (xv) we find

$$|e^{if(x')} - 1| \leq |e^{if(x')} - e^{if(x')} \hat{\lambda}(x')| + |\hat{\mu}(x') - 1|$$

$$= |1 - \hat{\lambda}(x')| + |\hat{\mu}(x') - 1| \leq 2\varepsilon.$$

If $x' \in F \cap C^o$ and $|t| \leq 1$, then $tx' \in F \cap C^o$, so if $\alpha = f(x')$ then $|e^{it\alpha} - 1| \leq 2\varepsilon \leq 1$ for all $t \in [-1,1]$. It is now a simple exercise to show that this implies that $|\alpha| \leq 2\varepsilon\sqrt{2}$. Thus $|f(x')| \leq 2\varepsilon\sqrt{2}$ for all $x' \in F \cap C^o$. Hence, f is $\Pi(E',E)$-continuous on F, and so by Hahn-Banach Theorem f admits a $\Pi(E',E)$-continuous extension to all of E'. But then 2.V.(2) shows, that there exists a vector $s \in E$ such that

$$f(x') = \langle x', s \rangle = \sum_{1}^{\infty} \langle x', m_j \rangle \qquad \forall x' \in F.$$

Hence, (4.1.2) and (4.1.3) are proved.

Applying (3.12) and (3.11) to the random sequence $(X_1-m_1, X_2-m_2, \ldots)$ and to the seminorm

$$q(x) = \sup_n \|\sum_1^n x_j\| \qquad \forall x = (x_j) \in E^\infty$$

we find that (4.1.4) - (4.1.7) are equivalent. Now note that the condition $\phi(2t) \leq K\phi(t)$ implies:

$$\phi(t+u) \leq K(\phi(t) + \phi(u))$$

So, if $\alpha = \phi(\|s\|)$ then

(xvi) $\qquad \phi(\|x-s\|) \leq K(\phi(\|x\|+\alpha)), \quad \phi(\|x\|) \leq K(\phi(\|x-s\|+\alpha)).$

And since $\mu = \mathcal{L}(R + s)$ we find that

(xvii) $\qquad E\,\phi(\|\sum_1^\infty (X_j-m_j)\|) = \int_E \phi(\|x-s\|)\,\mu(dx)$

Thus (xvi) and (xvii) show that (4.1.7) and (4.1.8) are equivalent.

Since (4.1.9) evidently implies (4.1.7), we only have to show that (4.1.4)-(4.1.8) imply (4.1.9). Let

$$M = \sup_n \|\sum_1^n (X_j-m_j)\|, \qquad M_o = \sup_{n,m} \|\sum_{j=n}^{n+m} (X_j-m_j)\|$$

$$V_n = \|\sum_{j=n}^\infty (X_j-m_j)\|.$$

Then evidently $V_n \leq M_o \leq 2M$, and $V_n \to 0$ a.s. by (4.1.2). Thus, if (4.1.5) holds, then $\phi(V_n) \to 0$ a.s., since $\phi(t) \to 0$ as $t \to 0$. Moreover, $\phi(V_n) \leq \phi(M_o)$ and $E\,\phi(M_o) \leq KE\,\phi(M) < \infty$. So, by Lebesque Dominated Convergence Theorem we have:

$$E\,\phi(V_n) \xrightarrow[n\to\infty]{} 0 \quad \text{and} \quad E\,\phi(V_n) < \infty \quad \forall n \geq 1 .$$

Thus (4.1.9) holds and the theorem is proved. □

REMARKS. If (X_n) is a sequence of independent E-valued random vectors, then any sequence of (non-random) vectors (m_j) satisfying (4.1.2) is called a <u>centering sequence</u> for (X_n), and ΣX_n is said to be <u>essentially summable</u>, if (X_n) admits at least one centering sequence. The proof of Theorem 4.1 shows that we have

(4.1) ΣX_n is essentially summable, if and only if the symmetrized series ΣX_n^s converges a.s.

It would be nice, if we could obtain a formula for a centering sequence for an essentially summable series ΣX_n. In the 1-dimensional case this can be found in [18, Theorem A,p.251]. In the general case I know nothing except that $m_j \equiv 0$ is a centering sequence for a series ΣX_n satisfying (4.1.1), in each of the following three cases:

(4.2) X_n is symmetric $\forall n \geq 1$

(4.3) $\exists K$ compact $\subseteq E$ $\exists a > 0$, so that $P(S_n \in K) \geq a$ $\forall n$

(4.4) ΣX_n converges in probability.

Moreover, if X_n is integrable and $a_n = E X_n$, then $m_j \equiv 0$ is a centering sequence for ΣX_n (still satisfying (4.1.1)) in each of the following three (equivalent) cases:

(4.5) $\sup_n E\| \sum_1^n X_j \| < \infty$ and $\{a_n \mid n \geq 1\}$ relatively compact

(4.6) $E \sup_n \| X_j \| < \infty$ and $\{a_n \mid n \geq 1\}$ relatively compact

(4.7) $\int_E \| x \| \mu(dx) < \infty$ and $\{a_n \mid n \geq 1\}$ relatively compact.

Notice that if $\{s_n\}$ converges in $\sigma(E,F)$ and $X_1 = s_1$, $X_n = s_n - s_{n-1}$ for $n \geq 2$, then (4.1.1) holds with μ equal to the Dirac measure at the limit of $\{s_n\}$. This shows that there is no hope for Σm_j to converge in any stronger topology than $\sigma_s(E,F)$, where $\sigma_s(E,F)$ is the strongest topology on E having the same convergent <u>sequences</u> (not nets!) as $\sigma(E,F)$. If $E = \ell^1$ and $F = \ell^\infty$, then $\sigma_s(\ell^1,\ell^\infty)$ equals to the $\|\cdot\|$-topology (see [11, Theorem IV.8.12]). If F is a Banach space and $E = F'$, then $\sigma_s(E,F) = \Pi(E,F)$ (see [17,p.272]). □

As noted in 2.VI.(6), the Bernoulli sequence (ε_j) exhibits all the randomness of a symmetric sequence (X_n). Let us consider the following two subsets of E^∞:

$$B_\varepsilon(E) = \{ (x_j) \in E^\infty \mid \{\sum_1^n \varepsilon_j x_j\} \text{ is a.s. bounded}\}$$

$$C_\varepsilon(E) = \{ (x_j) \in E^\infty \mid \sum_1^\infty \varepsilon_j x_j \text{ is a.s. convergent}\}.$$

By (3.11) we have that $(x_j) \in B_\varepsilon(E)$, respectively $(x_j) \in C_\varepsilon(E)$, if and only if

$$(4.8) \qquad \{\sum_1^n \varepsilon_j x_j\} \quad \text{is bounded in} \quad L_E^p(P) ,$$

respectively

$$(4.9) \qquad \sum_1^\infty \varepsilon_j x_j \quad \text{converges in} \quad L_E^p(P) ,$$

where p is any fixed number in the interval $[0,\infty[$. So, if we define $|x|_p$ by

$$|x|_p = \sup_n \|\sum_1^n \varepsilon_j x_j\|_p \qquad \forall x = (x_j) \in E^N .$$

Then $|x|_p < \infty$ for all $x \in B_\varepsilon(E)$ and all $0 \leq p < \infty$. It is then a matter of routine to prove the following theorem:

THEOREM 4.2. $(B_\varepsilon(E), |\cdot|_p)$ *is a Fréchet space for* $0 \leq p < 1$, *and a Banach space for* $1 \leq p < \infty$. *All the* $|\cdot|_p$*-topologies coincide for* $0 \leq p < \infty$, *and* $C_\varepsilon(E)$ *is the* $|\cdot|_p$*-closure of the space of finite sequences:*

$$(4.21) \qquad E^{(\infty)} = \{(x_j) \mid \exists n : x_j = 0 \qquad \forall j \geq n \}.$$

There exists a constant $K_{pq} > 0$, *such that*

$$(4.2.2) \qquad E\|\sum_1^n \varepsilon_j x_j\|^p \leq K_{pq} E\|\sum_1^n \varepsilon_j x_j\|^q$$

for all $n \geq 1$, *all* $x_1,\dots,x_n \in E$ *and all* $0 < p,q < \infty$. □

If $E = R$, then the inequalities (4.2.2) are known under the name "*the Khinchine inequalities*", and in the real valued case the best possible constants K_{pq} have recently been found by U.Haagerup (see [12]).

Now, note that if (X_n) is a symmetric sequence of E-valued random vectors, then by (2.56) we have

$$(4.10) \qquad P(\{\sum_1^n X_j\} \quad \text{is} \quad \text{bounded}) = P((X_j) \in B_\varepsilon(E))$$

$$(4.11) \qquad P(\sum_1^\infty X_j \quad \text{converges} \quad) = P((X_j) \in C_\varepsilon(E))$$

So, the two sets $B_\varepsilon(E)$ and $C_\varepsilon(E)$ give us the exact information about boundedness or convergence of partial sums of a symmetric sequence.

Let us introduce the following sequence spaces:

$$\ell^p(E) = \{x = (x_j) \in E^\infty \mid \|x\|_p = \{\sum_1^\infty \|x_j\|^p\}^{1/p} < \infty\},$$

$$\ell^\infty(E) = \{x = (x_j) \in E^\infty \mid \|x\|_\infty = \sup_j \|x_j\| < \infty\},$$

$$c_o(E) = \{(x_j) \in E^\infty \mid x_j \to 0\}.$$

Then $(\ell^p(E), \|\cdot\|_p)$ is a Banach space for all $1 \leq p \leq \infty$ and $c_o(E)$ is a closed linear subset of $\ell^\infty(E)$. It is clear that we have

$$\text{(4.12)} \qquad \ell^1(E) \subseteq C_\varepsilon(E) \subseteq B_\varepsilon(E) \subseteq \ell^\infty(E)$$

$$\text{(4.13)} \qquad C_\varepsilon(E) \subseteq c_o(E)$$

With these concepts at hand we can show the following striking connection between the geometry of a Banach space and a classical probabilistic limit theorem:

THEOREM 4.3. <u>The following five statements are mutually equivalent</u> :

(4.3.1) $B_\varepsilon(E) \subseteq c_o(E)$.

(4.3.2) $B_\varepsilon(E) = C_\varepsilon(E)$.

(4.3.3) <u>If</u> (X_n) <u>is symmetric and the set</u> $\{\sum_1^n X_j\}$ <u>is a.s. bounded, then</u> $\sum_1^\infty X_j$ <u>is a.s. convergent.</u>

(4.3.4) <u>If</u> $\{X_n\}$ <u>are independent and</u> $\{\sum_1^n X_j\}$ <u>is a.s. bounded, then</u> ΣX_j <u>is essentially convergent</u>.

(4.3.5) E <u>does not contain</u> c_o .

REMARK. $c_o = c_o(R)$, and (4.3.5) means, that no subspace of E is isomorphic to c_o .

PROOF. (4.3.1)$\Rightarrow$ (4.3.2): Suppose that (4.3.2) does not hold, i.e. $x = (x_j) \in B_\varepsilon(E) \setminus C_\varepsilon(E)$ for some x . Then $\sum_1^n \varepsilon_j x_j$ is <u>not</u> a Cauchy sequence in $L^1_E(P)$, so we can find integers $0 = \sigma(0) < \sigma(1) < \ldots,$

and $a > 0$, such that

$$E\|\sum_{\sigma(k-1) < j \leq \sigma(k)} \varepsilon_j x_j\| \geq a \qquad \forall k \geq 1 .$$

Now let X_k be the sum above. Then $\{X_k\}$ are independent, symmetric, and

$$M_o = \sup_k \|\sum_1^k X_k\| = \sup_k \|\sum_{j=1}^{\sigma(k)} \varepsilon_j x_j\|$$
$$\leq \sup_n \|\sum_1^n \varepsilon_j x_j\| = M$$

Since $(x_j) \in B_\varepsilon(E)$, we have that $M < \infty$ a.s., so by (4.10) we have that $(X_j) \in B_\varepsilon(E)$ a.s. Moreover, since $\|X_j\| \leq M_o$, $EM_o < \infty$ and $E\|X_j\| \geq a$ $\forall j$, it follows from Lebesque Dominated Convergence Theorem that $X_k(\omega) \not\to 0$ for all ω in a set of positive probability. That is

$$P((X_j) \in B_\varepsilon(E) \setminus c_o(E)) > 0$$

But this contradicts (4.3.1).

(4.3.2) $\Rightarrow$ (4.3.3) : Follows from (4.11) and (4.10).

(4.3.3) $\Rightarrow$ (4.3.4) : Follows from (4.1).

(4.3.4) $\Rightarrow$ (4.3.5) : Let T be a bounded linear map from c_o into E. We shall show, that T is not an isomorphism, and we shall do this by showing that $f_j = Te_j \to 0$ where e_j is the j-th unit vector in c_o, i.e. $e_j = (0,\ldots,0,1,0\ldots)$. Note that

$$\|\sum_1^n \varepsilon_j(\omega) f_j\| = \|T(\sum_1^n \varepsilon_j(\omega) e_j)\| \leq \|T\|$$

since

$$\|\sum_1^n \varepsilon_j(\omega) e_j\|_\infty = \max_{1\leq j\leq n} |\varepsilon_j(\omega)| = 1 .$$

Hence, the set $\{\sum_1^n \varepsilon_j f_j\}$ is a.s. bounded, and so $\Sigma \varepsilon_j f_j$ is essentially convergent by assumption. But a symmetric series is essentially convergent if and only if it is a.s. convergent. Thus $\Sigma \varepsilon_j f_j$ converges a.s. and so $(f_j) \in C_\varepsilon(E) \subseteq c_o(E)$.

(4.3.5) $\Rightarrow$ (4.3.1) : Let $(x_j) \in B_\varepsilon(E)$, then $X_j = \varepsilon_j x_j$ is a vector in $L_E^1(P)$, such that (X_j) has unconditionally bounded partial sums, i.e.

(i) $$\sup_n \sup_\pm \|\sum_1^n \pm X_j\|_1 = \sup_n \|\sum_1^n \varepsilon_j x_j\|_1 < \infty .$$

By a theorem of Kwapien ([16]), we have that $L^1_E(P)$ does not contain c_o whenever (4.3.5) holds. By a theorem of Bessaga and Pelczynski ([8]), we have that if $L^1_E(P)$ does not contain c_o, and (i) holds, then ΣX_j is unconditionally convergent in $L^1_E(P)$. In particular we have that $\| x_j\| = \| X_j\|_1 \to 0$, and so $(x_j) \varepsilon c_o(E)$. □

This theorem naturally leeds to the question, when $\ell^p(E) \subseteq B_\varepsilon(E) \subseteq \ell^q(E)$ or $\ell^p(E) \subseteq C_\varepsilon(E) \subseteq \ell^q(E)$. We say that E is of <u>type</u> p $(1 \leq p \leq 2)$, if one of following 4 mutually equivalent statements is satisfied

(4.14) $\ell^p(E) \subseteq B_\varepsilon(E)$.

(4.15) $\ell^p(E) \subseteq C_\varepsilon(E)$.

(4.16) $\exists K > 0 : E \| \sum_1^n \varepsilon_j x_j \|^p \leq K \sum_1^n \| x_j \|^p \quad \forall x_1, \ldots, x_n \varepsilon E$.

(4.17) $\exists K : E \| \sum_1^n X_j \|^p \leq K \sum_1^n E \| X_j \|^p \quad \forall X_1, \ldots, X_n$ independent E-valued random vectors with mean 0 .

We say E is of <u>cotype</u> q $(2 \leq q \leq \infty)$, if one of the following 4 conditions is satisfied

(4.18) $B_\varepsilon(E) \subseteq \ell^q(E)$.

(4.19) $C_\varepsilon(E) \subseteq \ell^q(E)$.

(4.20) $\exists k > 0 : E \| \sum_1^n \varepsilon_j x_j \|^q \geq k \sum_1^n \| x_j \|^q \quad \forall x_1, \ldots x_n \varepsilon E$.

(4.21) $\exists k > 0 : E \| \sum_1^n X_j \|^q \geq k \sum_1^n E \| X_j \|^q \quad \forall X_1, \ldots, X_n$ independent E-valued random vectors with mean 0 .

(NB: (4.20) and (4.21) only apply to the case $q < \infty$). By (4.12) we have that every Banach space is of type 1 and of cotype ∞ and since $B_\varepsilon(R) = C_\varepsilon(R) = \ell^2$, we have that the type of E belongs to $[1,2]$ and the cotype of E belongs to $[2, \infty]$, provided that $E \neq 0$.

The defintions of type and cotype make sense for any seminormed space $(E, \| \cdot \|)$ and we shall occasionally use them for such spaces.

We define <u>the type of</u> E , denoted p(E) , and <u>the cotype of</u> E , denoted q(E), by

$$p(E) = \sup \{ p \,\varepsilon [1,2] \mid E \text{ is of type } p \} ,$$

$$q(E) = \inf \{ q \,\varepsilon [2,\infty] \mid E \text{ is of type } q \} .$$

Clearly, we have:

(4.22) E is of type p for all $1 \leq p < p(E)$, and is not of type p for any $p > p(E)$.

(4.23) E is of cotype q for all $q(E) < q \leq \infty$, and is not of cotype q for any $q < q(E)$.

If E is of type p and $p = p(E)$ we say that E is of <u>exact type</u> p . And if E is of cotype q and $q = q(E)$ we say that E is of <u>exact cotype</u> q .

EXAMPLES 4.4. (i) Let $(S, \mathcal{B}, \mu)$ be a measure space such that $L^1(\mu)$ is infinite dimensional, and let $1 \leq \alpha < \infty$. Then $L^\alpha(\mu)$ is of exact type $p = \alpha \wedge 2$, and of exact cotype $q = \alpha \vee 2$.

To see this, we note that if $x_1, \ldots, x_n \in L^\alpha(\mu)$ then we have

$$E \|\sum_1^\infty \varepsilon_j x_j\|_\alpha^\alpha = \int_\Omega P(d\omega) \int_S |\sum_{j=1}^n \varepsilon_j(\omega) x_j(s)|^\alpha \mu(ds)$$

$$= \int_S E | \sum_1^n \varepsilon_j x_j(s)|^\alpha \mu(ds)$$

And by the Khinchine inequalities (real valued case) there exist constant K_α , $k_\alpha > 0$, such that

$$k_\alpha \{ \sum_1^n |x_j(s)|^2 \}^{\alpha/2} \leq E | \sum_1^n \varepsilon_j x_j(s)|^\alpha \leq K_\alpha \{ \sum_1^n |x_j(s)|^2 \}^{\alpha/2}$$

It easily follows from there, that $L^\alpha(\mu)$ is of type $p = \alpha \wedge 2$ and of cotype $q = \alpha \vee 2$.

On the other hand, since $L^1(\mu)$ is infinite dimensional there exist disjoint sets $B_1, B_2, \ldots \in \mathcal{B}$ with $0 < \mu(B_j) < \infty$. Now put

$$x_j(s) = \mu(B_j)^{-1/\alpha} 1_{B_j}(s)$$

Then

$$E \|\sum_{j=1}^n \varepsilon_j x_j\|_\alpha^r = n^{r/\alpha} \quad \text{and} \quad \sum_{j=1}^n \|x_j\|_\alpha^r = n$$

so $p(L^\alpha(\mu)) \leq \alpha \wedge 2$, and $q(L^\alpha(\mu)) \geq \alpha \vee 2$. Thus $\alpha \wedge 2$ and $\alpha \vee 2$ are the exact type and cotype of $L^\alpha(\mu)$.

(ii) Let $(S, \mathcal{B}, \mu)$ be a measure space, such that $L^\infty(\mu)$ is infinite dimensional, then $L^\infty(\mu)$ has exact type 1 and exact cotype ∞. To see this let $B_1, B_2, \ldots$ be disjoint sets in $\mathcal{B}$, so that $\mu(B_j) > 0$ for all $j \geq 1$. Put $x_j = 1_{B_j}$. Then

$$E\|\sum_1^n \varepsilon_j x_j\|_\infty^r = 1, \qquad \sum_1^n \|x_j\|_\infty^r = n,$$

so $q(L^\infty(\mu)) = \infty$. Now put

$$y_j(s) = \begin{cases} +1 & \text{if } s \in B_k \text{ and } k \equiv 0,1,\dots,2^{j-1}-1 \pmod{2^j} \\ -1 & \text{if } s \in B_k \text{ and } k \equiv 2^{j-1},\dots,2^j-1 \pmod{2^j} \\ 0 & \text{if } s \notin \bigcup_1^\infty B_k . \end{cases}$$

Let $\omega \in \Omega$, and put $\alpha_i = \frac{1}{2}(\varepsilon_{i+1}(\omega) + 1)$ and

$$k = \sum_{i=0}^n \alpha_i 2^i .$$

Then

$$k \equiv \sum_{i=0}^{j-1} \alpha_i 2^i \pmod{2^j} \qquad \forall j = 1,2,\dots,n .$$

Thus if $s \in B_k$, then $y_j(s) = (-1)^{\alpha_{j-1}} = \varepsilon_j(\omega)$, and so

$$\sum_{j=1}^n \varepsilon_j(\omega)\, y_j(s) = n .$$

Hence, we have

$$E\|\sum_1^n \varepsilon_j y_j\|_\infty^r = n^r \qquad \sum_1^n \|y_j\|_\infty^r = n$$

showing that $p(L^\infty(\mu)) = 1$.

(iii) If T is an infinite completely regular space, then $C(T)$ has the exact type 1 and the exact cotype ∞. This is shown exactly as in (ii).

(iv) Let $(S, \mathcal{B})$ be a measurable space, and $ca(S, \mathcal{B})$ be the set of all bounded signed measures on $(S, \mathcal{B})$ with its total variation norm :

$$\|\mu\|_1 = \sup \{ \sum_1^n |\mu(B_j)| : B_1,\dots,B_n \text{ disjoint } \in \mathcal{B} \}$$

If $\mathcal{B}$ is infinite, then $ca(S, \mathcal{B})$ has the exact type 1 and the exact cotype 2 . This shown as in (i), noticing that $ca(S, \mathcal{B})$ behaves like an L^1-space. □

The concept of type and cotype is closely related to the geometric shape of the finite dimensional sections of the unit ball of E. Let $1 \leq \lambda < \infty$, $1 \leq r < \infty$ and $n \in N$, and consider the following two propositions:

$P(\lambda,n,r)$: $\exists x_1,\ldots,x_n \in E$ so that $\forall t_1\ldots t_n \in R$ we have

$$\{\sum_{j=1}^{n} |t_j|^r\}^{1/r} \leq \|\sum_{j=1}^{n} t_j x_j\| \leq \lambda \sum_{j=1}^{n} |t_j|$$

$Q(\lambda,n,r)$: $\exists x_1,\ldots,x_n \in E$ so that $\forall t_1\ldots t_n \in R$ we have

$$\max_{1\leq j\leq n} |t_j| \leq \|\sum_1^n t_j x_j\| \leq \lambda \{\sum_{j=1}^{n} |t_j|^r\}^{1/r} .$$

Then $P(\lambda,n,r)$ states that there exists an n-dimensional section of the unit ball in E, which is contained in as n-dimensional ℓ^r-unit ball, and which contains an n-dimensional ℓ^1-ball of radius $1/\lambda$. The proposition $Q(\lambda,n,r)$ states that there exists an n-dimensional section of the unit ball in E, which is contained in an n-dimensional ℓ^∞-unit ball, and which contains an n-dimensional ℓ^r-ball of radius $1/\lambda$.

THEOREM 4.5. <u>With the notation introduced above, and with</u> $p,q \in [1,\infty]$, <u>we have</u>

(4.5.1) $\quad p \geq p(E) \iff \exists\lambda > 1 : P(\lambda,n,p)$ <u>holds</u> $\forall n \geq 1$

$\iff P(\lambda,n,p)$ <u>holds</u> $\forall\lambda > 1 \quad \forall n \geq 1$

(4.5.2) $\quad q \leq q(E) \iff \exists\lambda > 1 : Q(\lambda,n,q)$ holds $\forall n \geq 1$

$\iff Q(\lambda,n,q)$ <u>holds</u> $\forall n \geq 1 \quad \forall\lambda > 1$.

PROOF (4.5.1) : Suppose that $P(\lambda,n,p)$ holds for all $n \geq 1$ for some $\lambda > 1$. Choose $x_1^n,\ldots,x_n^n$ according to $P(\lambda,n,p)$. Then $\|x_j^n\| \leq \lambda$ and

$$n^{s/p} = \{\sum_{j=1}^{n} |\varepsilon_j(\omega)|^p\}^{s/p} \leq \|\sum_{j=1}^{n} \varepsilon_j(\omega)\, x_j^n\|^s ,$$

so we find

$$n^{s/p} \leq E\, \|\sum_{j=1}^{n} \varepsilon_j\, x_j^n\|^s \quad , \quad \sum_{j=1}^{n} \|x_j^n\|^s \leq \lambda^s\, n .$$

Hence, if E is of type s then $s \leqq p$, and so $p \geqq p(E)$.

The other implications are very difficult and I shall refer the reader to [20] for a proof.

(4.5.2): Suppose that $Q(\lambda,n,q)$ holds for all $n \geqq 1$ for some $\lambda > 1$. Choose $x_1^n,\ldots,x_n^n$ according to $Q(\lambda,n,q)$. Then as above we find

$$E \|\sum_{j=1}^{n} \varepsilon_j x_j^n\|^s \leqq \lambda^s n^{s/r} , \quad \sum_{j=1}^{n} \|x_j^n\|^s \geqq n$$

so as before we conclude, that $q \leqq q(E)$.

Again the other implications are very difficult, and I shall refer the reader to [20] for a proof. □

Let (η_n) be a sequence of independent, identically distributed random variables, then we can define

$$B_\eta(E) = \{(x_j) \in E^\infty \mid \{\sum_1^n \eta_j x_j\} \text{ is a.s. bounded}\} ,$$

$$C_\eta(E) = \{(x_j) \in E^\infty \mid \sum_1^\infty \eta_j x_j \text{ is a.s. convergent}\} .$$

It can then be shown that if η_1 is non-degenerated (i.e. not constant a.s.) then (see [4])

(4.24) $$B_\eta(E) \subseteq B_\varepsilon(E) \quad \text{and} \quad C_\eta(E) \subseteq C_\varepsilon(E)$$

Moreover, if $E\eta_1 = 0$ and $0 < E|\eta_1|^p < \infty$ for some $p \in [1,2]$, then (see [4]) :

(4.25) $$E \text{ is of type } p \iff \ell^p(E) \subseteq B_\eta(E) \iff \ell^p(E) \subseteq C_\eta(E) .$$

And if $E\eta_1 = 0$ and $0 < E|\eta_1|^{q+\varepsilon} < \infty$ for some $q \in [2,\infty[$ and some $\varepsilon > 0$, then (see [4])

(4.26) $$E \text{ is of cotype } q \iff \ell^q(E) \supseteq B_\eta(E) \iff \ell^q(E) \supseteq C_\eta(E) .$$

Thus we may use any sequence of independent identically distributed random variables (η_j) with $E\eta_1 = 0$ and $E|\eta_1|^r < \infty \,\forall\, 0 < r < \infty$, in the definition (4.14)-(4.15) of type p, and in the definition (4.18)-(4.19) of cotype q. In particular, we may use a sequence of independent $N(0,1)$- distributed random variables (γ_j) instead of the Bernoulli sequence.

5. THE LAW OF LARGE NUMBERS

In this section we shall study the validity of the three versions of the law of large numbers given in (1.2)-(1.4). The first holds true in full generality:

THEOREM 5.1. Let $(E,\|\cdot\|)$ be a Banach space and $\{X_n\}$ a sequence of independent identically distributed E-valued random variables with mean $m = E\,X_1$. Then we have

(5.1.1) $$\frac{1}{n}\sum_{j=1}^{n} X_j \to m \quad \text{a.s. in } \|\cdot\| .$$

PROOF. Let $\mu = \mathcal{L}(X_1)$, then $f(x) = x$ belongs to $L^1_E(\mu)$, and so if $\varepsilon > 0$ is given then there exists a simple function $f_o : E \to E$ such that

(i) $$\int_E \|f_o(x) - x\| \,\mu(dx) \leq \varepsilon ,$$

(ii) $$\int_E f_o(x)\,\mu(dx) = m .$$

Put $Y_n = f_o(X_n)$, then $\{Y_n\}$ are independent identically distributed E_o-valued random variables, where E_o is the closed linear span of the range $f_o(E)$ of f_o. Since f_o is simple (i.e. takes only finitely many values) we have that E_o is finite dimensional. Thus by the finite dimensional law of large numbers we have (note that $m = E\,Y_n$ by (ii)):

(iii) $$\frac{1}{n}\sum_{j=1}^{n} Y_j \to m \qquad \text{a.s.}$$

Thus we find:

(iv) $$\|m - \frac{1}{n}\sum_{j=1}^{n} X_j\| \leq \|m - \frac{1}{n}\sum_{j=1}^{n} Y_j\| + \frac{1}{n}\sum_{j=1}^{n}\|X_j - Y_j\| .$$

Now, notice that $\xi_j = \|X_j - Y_j\|$ $(j=1,2,\dots)$ are independent identically distributed (real) random variables with

(v) $$E\,\xi_j = E\|X_j - Y_j\| = \int_E \|x - f_o(x)\|\,\mu(dx) \leq \varepsilon .$$

So by (iii) and the real valued law of large numbers we have

$$\limsup_n \| m - \frac{1}{n} \sum_1^n X_j(\omega) \| \leq \int_E \| x - f_o(x) \| \ \mu(dx) \leq \varepsilon$$

for a.a. $\omega \in \Omega$. Since $\varepsilon > 0$ is arbitrary, (5.1.1) follows.□

In order to discuss the second law of large numbers, viz.(1.3) we need a certain convexity condition. A Banach space E is said to be B-convex if there exist a number $0 < \lambda < 1$ and an integer $k \geq 2$, so that the following condition holds

$B(\lambda,k)$: $\forall x_1,\ldots x_k \in B \quad \exists \alpha_1,\ldots,\alpha_k \in \{+1,-1\}$, so that

$$\| \sum_{j=1}^{k} \alpha_j x_j \| \leq \lambda k ,$$

where $B = \{x \in E \mid \| x \| \leq 1\}$ is the unit ball in E .

EXAMPLES 5.2. (i): If $(E,\|\cdot\|)$ is a Hilbert space and $x,y \in B$, then by the parallelogram law we have:

$$\| x+y \|^2 + \| x-y \|^2 = 2(\| x \|^2 + \| y \|^2) \leq 4$$

So either $\| x+y \| \leq \sqrt{2}$ or $\| x-y \| \leq \sqrt{2}$. Thus $B(\sqrt{\frac{1}{2}},2)$ holds and so every Hilbert space is B-convex.

(ii) Let $E = L^p(\mu)$ where $1 \leq p \leq \infty$. If $1 \leq p \leq 2$ we have the following form of the parallelogram law:

$$\| f+g \|_p^p + \| f-g \|_p^p \leq 2(\| f \|_p^p + \| g \|_p^p)$$

So, if $f,g \in B$ then either $\| f+g \|_p \leq 2^{1/p}$ or $\| f-g \|_p \leq 2^{1/p}$. Thus, $B(2^{1/p-1},2)$ holds true.

If $2 \leq p < \infty$, we have the following form of the parallelogram law:

$$\| f+g \|_p^p + \| f-g \|_p^p \leq 2^{p-1} (\| f \|_p^p + \| g \|_p^p)$$

As above we find that $B(2^{-1/p},2)$ holds true.

Thus, we have shown that $L^p(\mu)$ is B-convex whenever $1 < p < \infty$. Inspecting Example 4.4. one finds that $L^1(\mu)$, $L^\infty(\mu)$, $C(T)$ and $ca(S,\mathcal{B})$ are not B-convex provided that they are infinite dimensional.

THEOREM 5.3. Let $(E,\|\cdot\|)$ be a Banach space, then the following four statements are mutually equivalent:

(5.3.1) E is B-convex

(5.3.2) $p(E) > 1$

(5.3.3) $\frac{1}{n}\sum_{1}^{n} X_j \to 0$ a.s., whenever $\{X_j\}$ are independent, E-valued, $EX_j = 0$ and $\sup_j E\|X_j\|^2 < \infty$.

(5.3.4) $\frac{1}{n}\sum_{1}^{n} \varepsilon_j x_j \to 0$ in probability $\forall (x_j) \in \ell^\infty(E)$.

PROOF. The implication : (5.3.1)$\Rightarrow$(5.3.2) is rather involved, and I shall refer the reader to [4, p.121-126].

(5.3.2) $\Rightarrow$ (5.3.3) : Let $1 < p < p(E)$, then we have

$$E\|\sum_{j=n+1}^{n+m} j^{-1} X_j\|^p \le K \sum_{j=n+1}^{n+m} j^{-p} E\|X_j\|^p$$

$$\le K \sum_{j=n+1}^{\infty} j^{-p} (E\|X_j\|^2)^{p/2} \le KC \sum_{j=n+1}^{\infty} j^{-p}$$

where $C = (\sup_j E\|X_j\|^2)^{p/2}$ and K is the constant appearing in (4.17). Thus $\sum_1^\infty j^{-1} X_j$ converges in $L_E^p(P)$, and so it converges a.s. by Theorem 4.1 and (4. 4). Hence, by the Kronecker lemma:

(i) $$\sum_{j=1}^{\infty} j^{-1} x_j \text{ converges} \Rightarrow \frac{1}{n}\sum_{1}^{n} x_j \to 0 ,$$

we find that $\frac{1}{n}\sum_{1}^{n} X_j \to 0$ a.s.

(5.3.3) $\Rightarrow$ (5.3.4) : Evident!

(5.3.4) $\Rightarrow$ (5.3.1) : Suppose that E is not B-convex. Then we can find $x_j \in B$ so that

$$\|\sum_{3^k < j \le 3^{k+1}} \alpha_j x_j\| \ge \frac{3}{4}(3^{k+1} - 3^k)$$

for all $k = 0,1,2,\ldots$ and all $\alpha_j = \pm 1$. Let (ε_j) be a Bernoulli sequence. Then

$$\|\sum_{j=1}^{3^k} \varepsilon_j(\omega) x_j\| \ge \|\sum_{3^{k-1} < j \le 3^k} \varepsilon_j(\omega) x_j\| - \sum_{1 \le j \le 3^{k-1}} \|x_j\|$$

$$\ge \frac{3}{4}(3^k - 3^{k-1}) - 3^{k-1} = \frac{1}{2}\, 3^{k-1}$$

Hence

$$\| \frac{1}{n} \sum_{j=1}^{n} \varepsilon_j(\omega)\, x_j \| \geq \frac{1}{6} \qquad \forall \omega \in \Omega \ \forall n=1,3,9,27,\ldots ,$$

so $\frac{1}{n} \sum_1^n \varepsilon_j x_j \not\to 0$ in probability. But $x = (x_j) \in \ell^\infty(E)$ and so we have derived a contradiction to (5.3.4).□

THEOREM 5.4. *Let* E *be a Banach space and* $p \in [1,2]$ *, then the following three statements are equivalent:*

(5.4.1) E *is of type* p

(5.4.2) $\frac{1}{n} \sum_1^n X_j \to 0$ *a.s., whenever* $\{X_j\}$ *is independent, E-valued,* $E X_j = 0$ *and* $\Sigma_1^\infty j^{-p} E\|X_j\|^p < \infty$

(5.4.3) $\frac{1}{n} \sum_1^n \varepsilon_j x_j \to 0$ *in probability for all* $x = (x_j) \in E^\infty$ *satisfying* : $\Sigma_1^\infty j^{-p} \|x_j\|^p < \infty$.

PROOF. (5.4.1) ⟹ (5.4.2): This is done exactly as the proof of "(5.3.2) ⟹ (5.3.3)".

(5.4.2) ⟹ (5.4.3) : Evident!

(5.4.3) ⟹ (5.4.1) : Let us define two norms on E^∞

$$\|x\|_o = \{ \sum_1^\infty j^{-p} \|x_j\|^p \}^{1/p} , \quad G_o = \{ x \mid \|x\|_o < \infty \}$$

$$\|x\|_1 = \sup_n \| \frac{1}{n} \sum_1^n \varepsilon_j x_j \|_p , \quad G_1 = \{ x \mid \|x\|_1 < \infty \}$$

It is then a matter of routine to verify, that $(G_j, \|\cdot\|_j)$ is a Banach space for $j = 0,1$. Since $|\cdot|_o$ and $|\cdot|_p$ define the same topology on $B_\varepsilon(E)$ (see Theorem 4.2), we see that (5.4.3) implies that $G_o \subseteq G_1$. The injection $G_o \to G_1$ has a closed graph, so by the closed graph theorem there exists a constant K , such that

$$E \| \sum_{j=1}^n \varepsilon_j x_j \|^p \leq K n^p \sum_{j=1}^n j^{-p} \|x_j\|$$

for all $n \geq 1$ and all $x_1,\ldots,x_n$. Applying this to the vector $(0,\ldots,0,x_1,\ldots,x_n)$ where the k first coordinates are 0, we obtain

$$E\|\sum_{1}^{n} \epsilon_j x_j\|^p \leq K(k+n)^p \sum_{j=1}^{n} (j+k)^{-p}\|x_j\|^p$$

$$\leq K(\frac{k+n}{k+1})^p \sum_{j=1}^{n} \|x_j\|^p$$

Letting $k \to \infty$ we see that (4.16) is satisfied and so E is of type p .□

It is interesting to note that (5.3.3) and (5.3.4) are isomorphic properties of E , i.e. independent of the particular choice of the norm. Thus, B-convexity is an isomorphic property, i.e. if $\|\cdot\|_o$ and $\|\cdot\|_1$ are two equivalent norms on E , such that $(E,\|\cdot\|_o)$ is B-convex, then so is $(E,\|\cdot\|_1)$. This is not clear from the definition and in general both the number λ and the integer k appearing in $B(\lambda,k)$ may be different for the two norms.

EXAMPLE 5.5. Let me describe a typical case to which Theorem 5.1 applies (sometimes!).

Let $(S, \mathcal{B}, \mu)$ be a probability space, Θ a set, and $(E_o,\|\cdot\|)$ a normed linear subspace of R^Θ with the completion $(E,\|\cdot\|)$. Suppose that $E_1 \subseteq E_o$ is a functionally closed subset of E (see VII(8)). Let g be a map : $S \times \Theta \to R$, and $Y_1, Y_2, \ldots$ a sequence of independent S-valued random variables with $\mathcal{L}(Y_n) = \mu$ $n=1,2,\ldots$. Now suppose that

(5.1) $$G(s) = g(s,\cdot) \in E_1 \quad \forall s \quad \text{and} \quad G \in L^1_E(P).$$

That is $\{g(\cdot,\theta) | \theta \in \Theta\}$ is an E-valued and E-integrable process defined on $(S, \mathcal{B}, \mu)$ (see(2.69)-(2.71)).Let m be the mean function of g :

$$m(\theta) = \int_S g(s,\theta) \quad \mu(ds)$$

then $X_n = G(Y_n)$ $(n=1,2,\ldots)$ are independent identically distributed E-valued random vectors with $E X = m$ (see (2.72)). Thus by Theorem 5.1 we have

(5.2) $$\frac{1}{n} \sum_{j=1}^{n} g(Y_j, \theta) \xrightarrow[n\to\infty]{} m(\theta) \quad \text{a.s. in} \quad (E,\|\cdot\|).$$

Let us consider some special cases:

1° Bounded stochastic processes. Let $E = B(\Theta)$ with its sup-norm $\|\cdot\|_\infty$ (see 2.I.(4)). Then $(B(\Theta), \|\cdot\|_\infty)$ is a Banach subspace of R^Θ satisfying (2.73) so $B(\Theta)$ is functionally closed. Moreover, the condition (5.1) is equivalent to:

(5.3) $\quad g(s,\cdot)$ is bounded $\forall s \in S$,

(5.4) $\quad g(\cdot,\theta)$ is measurable $\forall \theta \in \Theta$,

(5.5) $\quad \rho(s,t) = \sup_\theta |g(s,\theta) - g(t,\theta)|$ is a separable pseudometric,

(5.6) $\quad \int_S \sup_\theta |g(s,\theta)| \ \mu(ds) < \infty$.

And if (5.3)-(5.6) holds, then we have

(5.7) $$\frac{1}{n}\sum_{j=1}^{n} g(Y_j,\theta) \xrightarrow[n\to\infty]{} m(\theta) \quad \text{uniformly in } \theta \text{ a.s.}$$

Condition (5.5) (i.e. separability of the range of X_j) is crucial for the proof of Theorem 5.1, and as we shall see below it may fail in important examples, where (5.7) can be proved by other methods.

2° Processes with continuous sample paths. Let Θ be a compact metric space. Then $(C(\Theta), \|\cdot\|_\infty)$ is a functionally closed Banach space, and (5.1) is equivalent to

(5.8) $\quad g(s,\cdot)$ is continuous $\forall s \in S$,

(5.9) $\quad g(\cdot,\theta)$ is measurable $\forall \theta \in \Theta$,

(5.10) $\quad \int_S \sup_\Theta |g(s,\theta)| \ \mu(ds) < \infty$.

And if (5.8)-(5.10) holds, then so does (5.7).

This takes care of processes with continuous sample paths and with a compact metric space as the time space. However, in applications we need more general time space, e.g. R_+ or R^n or an infinite dimensional vector space. So suppose that Θ is a polish space (complete separable metric space), and let $E = C(\Theta)$ (see 2.I.(3)) equipped with the topology K of uniform convergence on compact subsets of Θ, i.e. $\phi_\alpha \to \phi$ in K, if and only if $\|\phi_\alpha - \phi\|_\Gamma \to 0$ for all compact sets $\Gamma \subseteq \Theta$, where

$$\|\phi\|_\Gamma = \sup_{\theta\in\Gamma} |\phi(\theta)| \qquad \forall\Gamma \subseteq \Theta \ \forall\phi \in C(\Theta).$$

Then $(E, \mathcal{K})$ is a locally convex vector space, but $(E, \mathcal{K})$ is not a Banach space unless Θ is compact. Suppose that g satisfies (5.8), (5.9) and

(5.11) $$\int_S \sup_{\theta\in\Gamma} |g(s,\theta)| \ \mu(ds) < \infty \qquad \forall\Gamma \text{ compact} \subseteq \Theta \ .$$

Due to the simple lattice structure of $\mathcal{K}(\Theta)$ (=the set of all compact subsets of Θ, see [9,p.58]) it can be shown (see [14]) that we have

(5.12) $$\frac{1}{n}\sum_1^n g(Y_j, \Theta) \xrightarrow[n\to\infty]{} m(\theta) \qquad \text{in} \quad \mathcal{K} \qquad \text{a.s.}$$

3° p-th order stochastic processes. Let $(\Theta, \mathcal{A}, \nu)$ be a measure space, and $1 \leq p < \infty$. Then $(L^p(\nu), \|\cdot\|_p)$ is a functionally closed Banach space, and (5.1) is equivalent to

(5.13) $\qquad g$ is $\mathcal{B}\otimes\mathcal{A}$-measurable

(5.14) $$\int_S \{ \int_\Theta |g(s,\theta)|^p \nu(d\theta) \}^{1/p} \mu(ds) < \infty \ .$$

And if (5.13)-(5.14) holds then we have

(5.15) $$\int_\Theta | m(\theta) - \frac{1}{n}\sum_{j=1}^n g(Y_j,\Theta)|^p \ \nu(d\theta) \xrightarrow[n\to\infty]{} 0 \quad \text{a.s.}$$

4° Parametric statistics. In parametric statistics one considers the following problem:

Let $Y_1, Y_2, \ldots$ be independent, identically distributed S-valued random variables whose common distribution belongs to a parametrized family $\{\mu_\theta | \theta\in\Theta\}$ of probability measures on $(S, \mathcal{B})$. Let us also assume that μ_θ is absolutely continuous with respect to some fixed σ-finite measure λ on $(S, \mathcal{B})$ for all $\theta\in\Theta$, and let $f(s,\theta)$ denote the λ-density of μ_θ:

(5.16) $$f(s,\theta) \geq 0 \qquad \forall(s,\theta) \in S\times\Theta$$

(5.17) $$f(\cdot,\theta) \text{ is measurable} \quad \forall\theta\in\Theta$$

(5.18) $$\int_S f(s,\theta)\ \lambda(ds) = 1 \qquad \forall\theta\in\Theta$$

$$(5.19) \qquad \mu_\theta(B) = \int_B f(s,\theta)\,\lambda(ds) \qquad \forall\theta\in\Theta,\ \forall B\in\mathcal{B}$$

Let $\bar\theta$ be the "true but unknown" parameter value, i.e. suppose that $\mathcal{L}(Y_j) = \mu_{\bar\theta}$. In order to estimate $\bar\theta$, based on the observations, $Y_1,\ldots,Y_n$, we form the <u>likelihood function</u> :

$$(5.20) \qquad L_n(\omega,\theta) = \prod_{j=1}^{n} f(Y_j(\omega),\theta)$$

and the <u>log-likelihood function</u>:

$$(5.21) \qquad \ell_n(\omega,\theta) = \log L_n(\omega,\theta) = \sum_{j=1}^{n} \log f(Y_j(\omega),\theta)\ .$$

Then <u>the maximum likelihood estimator</u> of $\bar\theta$ is a random variable $\hat\theta_n(\omega)$, such that $L_n(\omega,\cdot)$, or equivalently $\ell_n(\omega,\cdot)$, assumes its maximum for $\theta = \hat\theta_n(\omega)$.

Put $g(s,\theta) = \log f(s,\theta)$, then the log-likelihood function is given by

$$(5.22) \qquad \ell_n(\theta) = \sum_{j=1}^{n} g(Y_j,\theta)$$

and the mean function m is nothing but the so called <u>Fisher's entropy function</u> $H(\theta|\bar\theta)$ where

$$(5.23) \qquad H(\theta_1|\theta_2) = \int_S f(s_1,\theta_1)\ \log f(s,\theta_2)\,\lambda(ds)$$

(i.e. $m(\theta) = H(\theta|\bar\theta)$). It is easily checked that we have $H(\cdot|\bar\theta)$ assumes its maximum at $\theta = \bar\theta$, and if the problem is <u>well posed</u> (i.e. $\mu_\theta \neq \mu_{\bar\theta}$ for all $\theta\neq\bar\theta$), then $\bar\theta$ is a unique maximum for $H(\cdot|\bar\theta)$.

Thus, under suitable condition on g (see 1° or 2°) we have that $\frac{1}{n}\ell_n(\theta) \to H(\theta|\bar\theta)$ uniformly on Θ (or on all compact subsets of Θ) with probability 1. This can be used to prove the consistency of the maximum likelihood estimator $\hat\theta_n$ (i.e. to show that $\hat\theta_n \to \bar\theta$ a.s.).

If Θ is an open subset of R^m, then one can find the maximum liklihood estimator $\hat\theta_n(\omega)$, by solving the likelihood equations:

$$(5.24) \qquad \frac{\partial}{\partial\theta_\nu}\ \ell_n(\omega,\theta) = 0 \qquad \text{for } \nu = 1,\ldots,m\ .$$

Or, equivalently (put $g_\nu(s,\theta) = \frac{\partial}{\partial\theta_\nu} g(s,\theta)$) :

$$(5.25) \qquad \sum_{j=1}^{n} g_\nu (Y_j(\omega), \theta) = 0 \quad \text{for } \nu = 1,2,\dots,m .$$

This leads to the study of the sums:

$$(5.26) \qquad \frac{\partial}{\partial \theta_\nu} \ell_n(\theta) = \sum_{1}^{n} g_\nu (Y_j, \theta) ,$$

which are sums of the form studied above. Note that the mean function $m_\nu(\theta)$ of g_ν takes the form

$$(5.27) \qquad m_\nu(\theta) = \int_S \frac{f(s,\bar\theta)}{f(s,\theta)} \frac{\partial}{\partial \theta_\nu} f(s,\theta)\lambda(ds) .$$

So under suitable regularity condition we have

$$(5.28) \qquad m_\nu(\theta) = \frac{\partial}{\partial \theta_\nu} H(\theta | \bar\theta) \qquad \forall \nu = 1,\dots,m$$

$$(5.29) \qquad m_\nu(\bar\theta) = \frac{\partial}{\partial \theta_\nu} \int_S f(s,\theta)\lambda(ds) \Big|_{\theta=\bar\theta} = 0 \ \forall \nu = 1,\dots,m .$$

Again the law of large numbers (Theorem 5.1) can be used to show that a solution $\tilde\theta_n$ to the likelihood equation (5.24) tends a.s. to $\bar\theta$ as $n \to \infty$, provided certain not too restrictive conditions are satisfied.

5° Empirical distribution functions. Let $S = \Theta = R$ and let $F(x) = P(Y_j \leq x)$ be the distribution function of Y_j. Put

$$(5.30) \qquad g(s,\theta) = \begin{cases} 1 & \text{if } s \leq \theta \\ 0 & \text{if } s > \theta . \end{cases}$$

Then

$$(5.31) \qquad \frac{1}{n} \sum_{j=1}^{n} g(Y_j(\omega),x) = \frac{1}{n} \sum_{j=1}^{n} 1_{]-\infty,x]}(Y_j(\omega)) = F_n(\omega,x)$$

is the empirical distribution function based on the observations $Y_1,\dots,Y_n$, and the mean function equals the distribution function: $m(x) = F(x)$. It is well-known (see [18, p.20]) that we have (the Glivenko-Cantelli theorem):

$$(5.32) \qquad F_n(\omega,x) \xrightarrow[n\to\infty]{} F(x) \quad \text{uniformly in } x \in R \qquad \text{a.s.}$$

The function g satisfies (5.3), (5.4) and (5.6), but g does not satisfy (5.5) (note that $\rho(s,t) = 1$ if $s \neq t$ and $\rho(s,s) = 0$). Hence Theorem 5.1 does not apply to this case.

It would be interesting to have a vector-valued law of large numbers, which would include (5.32) as a special case, since it may throw a new light to the so called Glivenko classes (see below).

6° Empirical laws. Let Θ be a subset of $\mathcal{B}$, where $\mathcal{B}$ is the σ-algebra on S, and put

$$(5.33) \qquad g(s,\theta) = 1_\theta(s)$$

Then

$$\frac{1}{n}\sum_{j=1}^{n} g(Y_j(\omega),\theta) = \frac{1}{n}\sum_{j=1}^{n} 1_\theta(Y_j(\theta)) = \mu_n(\omega,\theta)$$

is the empirical law based on $Y_1,\ldots,Y_n$, and the mean function m equals μ, i.e. $m(\theta) = \mu(\theta)$ for all $\theta \in \Theta$. The function g satisfies (5.3), (5.4) and (5.6), but g does not satisfy (5.5) in general. If we have

$$(5.34) \qquad \mu_n(\omega,\theta) \xrightarrow[n\to\infty]{} \mu(\theta) \text{ uniformly in } \theta \text{ a.s.}$$

then Θ is called a Glivenko class of μ. E.g. if $S = R^n$ and μ is absolutely continuous with respect to Lebesque measure, then the set of all convex subsets of R^n is a Glivenko class of μ (see [21], where more information about the subject can be found). Again it would be interesting to have a general vector-valued law of large numbers, which would cover some of the known Glivenko classes.

7° Random measures. Let (T,ρ) be a separable metric space and let $\Theta = \mathcal{B}(T)$. In 2.II(6)-(8) we introduced the Lipshitz' function space $\mathrm{Lip}(S,\rho)$ and its dual $M(S,\rho)$, and we saw that $ca_\tau(T)$ was a separable linear subspace of $(M(S,\rho),\|\cdot\|_\rho)$ (see in particular 2. II.(8)). So let $(E_0,\|\cdot\|) = (ca_\tau(T),\|\cdot\|_\rho)$, then $(E_0,\|\cdot\|)$ is a normed separable subspace of R^Θ. So, let E be the completion of E_0, i.e. E is the $\|\cdot\|_\rho$-closure of E_0 in $M(T,\rho)$. Then $E_1 = ca_\tau^+(T)$ is functionally closed in E (see 2.VII.(7)-(9)). Hence, if g is a positive, μ-integrable, measure kernel on $S \times T$, i.e. if

(5.35) $\quad g(s,\cdot)$ is a finite positive Borel measure on T ,

(5.36) $\quad g(\cdot,B)$ is measurable $\forall B \in \mathcal{B}(T)$,

(5.37) $\quad \int_S g(s,T)\, \mu(ds) < \infty$,

Then $G(s) = g(s,\cdot)$ is an integrable E_1-valued random vector and $m = \int G\, d\mu$ is the <u>mixture</u> of g with μ , i.e. m is the measure defined by

(5.38)
$$m(B) = \int_S g(s,B)\, \mu(ds) \quad \forall B \in \mathcal{B}(T) .$$

By Theorem 5.1 and (2.14) we have

(5.39)
$$\frac{1}{n} \sum_{j=1}^{n} g(Y_j,\cdot) \xrightarrow[n \to \infty]{\sim} m(\cdot) \qquad \text{a.s.}$$

In particular we find

(5.40)
$$\mu_n(\omega,\cdot) \overset{\sim}{\to} \mu \text{ for } P\text{- a.a. } \omega \in \Omega ,$$

where μ_n is the empirical law of μ based on $Y_1,\ldots,Y_n$. □

6. THE CENTRAL LIMIT THEOREM

In this section we shall study the central limit theorem in a Banach space $(E, \|\cdot\|)$. So let $X_1, X_2, \dots$ be independent, identically distributed, E-valued, random vectors with $E\, X_n = 0$ and $E\|X_n\|^2 < \infty$, and put

$$(6.1) \qquad U_n = \frac{1}{\sqrt{n}} \sum_{j=1}^{n} X_j \quad .$$

We can then ask, if $\{U_n\}$ converges in law to a centered gaussian measure γ on E (see 2.VII (5))? Let us analyze the question.

By the central limit theorem for real valued random variables we have that $\langle x', U_n \rangle \overset{\sim}{\to} N(0, r(x',x'))$ for all $x' \in E'$, where r is the covariance of X_1. Hence, we have

$$(6.2) \qquad E\, e^{i\langle x', U_n\rangle} \xrightarrow[n \to \infty]{} e^{-\frac{1}{2} r(x',x')} \qquad \forall x' \in E' .$$

Hence if $U_n \overset{\sim}{\to} \gamma$ for some Radon measure γ, then γ is necessarily a centered gaussian measure having r as its covariance, and so r must be a gaussian quadratic form on $E' \times E'$.

A random vector X is called pregaussian if its covariance r_X is gaussian, i.e. if $r_X = r_\gamma$ for some centered gaussian measure γ on E. So one requirement for a central limit theorem is that X_1 should be pregaussian.

Now, suppose that X_1 is pregaussian and let γ be a centered gaussian measure with the same covariance as X_1. Then by (6.2) and (2.47) we have

$$(6.3) \qquad U_n \overset{\sim}{\to} \gamma \quad \text{in} \quad (E, \sigma(E,E')) .$$

But we want convergence in law in the space $(E, \|\cdot\|)$, which means much more (there are many more $\|\cdot\|$-continuous real functions, than weakly continuous real functions). If X_1 is pregaussian and $U_n \overset{\sim}{\to} \gamma$ in $(E, \|\cdot\|)$, we shall say that X_1 belongs to <u>the domain of normal attraction</u>. The set of all E-valued random variables, which belong to the domain of normal attracion is denoted DNA.

Our next theorem shows that the answer to all our problems is "type 2 spaces":

THEOREM 6.1. Let $(E,\|\cdot\|)$ be a Banach space, then the following four statements are mutually equivalent

(6.1.1) X is pregaussian $\forall X \in L^2_E(P)$

(6.1.2) X is pregaussian $\forall X$ symmetric, discrete and satisfying $:\|X(\omega)\| = 1 \quad \forall \omega$

(6.1.3) E is of type 2

(6.1.4) $\frac{1}{\sqrt{n}} \sum_{j=1}^{n} X_j$ converges in law in $(E,\|\cdot\|)$ to a centered gaussian measure, whenever $X_1, X_2, \ldots$ are independent identically distributed random vectors with mean 0 and finite second moment.

P.s. A random vector is said to be discrete (simple) if it only assume countably many (finitely many) different values.

REMARK. I shall below only sketch the proof of the most important implication, viz. (6.1.3)=>(6.1.4). However, I shall sketch the two known proofs of this implication, but let me first make a few comments on the proof. In view of (6.2) it suffices to show that $\{\mathcal{L}(U_n) \mid n \geq 1\}$ is uniformly tight (see § 1), and the search for a Banach space valued central limit theorem was for a long time deluded in an attempt to prove this. None of the two proofs below applies this method, which is a standard method in probability theory, and both proofs introduce a new idea.

First proof of (6.1.3) =>(6.1.4). Let $\mu = \mathcal{L}(X_1)$, and let $\{W(B) \mid B \in \mathcal{B}(E)\}$ be a white noise with covariance measure μ, i.e. $\{W(B) \mid B \in \mathcal{B}(E)\}$ is a real valued stochastic process satisfying:

(i) $(W(A_1),\ldots,W(A_n))$ has a centered n-dimensional gaussian distribution $\forall A_1,\ldots,A_m \in \mathcal{B}(E)$

(ii) $E\{W(A)\, W(B)\} = \mu(A \cap B) \quad \forall A,B \in \mathcal{B}(E)$

It is then easily checked, that $W(\cdot)$ is an $L^2(P)$-valued vector measure on $(E, \mathcal{B}(E))$. Then one introduces an E-valued stochastic integral with respect to W, i.e. one defines:

$$\int_E f(x)\ W(dx)$$

for a suitable class, denoted by $L_E^2(W)$, of functions from E into E . If f is a simple function, the definition of $\int f\ d\ W$ is evident; one closes the set of simple functions with respect to the norm

$$\|f\|_W = \{ \int_E \| f(x)\|^2 \mu\ (dx) + E \| \int_E f(x)\ W(dx)\|^2 \}^{1/2}$$

Thus, $L_E^2(W)$ becomes a subset of $L_E^2(\mu)$; actually it turns out that $f \in L_E^2(W)$, if and only if $f \in L_E^2(\mu)$ and the covariance of f

$$\sigma_f(x',y') = \int_E <x',x> <y',x> \mu(dx) \quad \text{for } x',y' \in E'$$

is gaussian. Moreover, in the case we have

(iii) $\int_E f\ d\ W$ has a centered gaussian distribution law with covariance σ_f .

Then one shows fairly easy that if E is of type 2 , then $L_E^2(\mu)=L_E^2(W)$, and there exists a constant $K > 0$ such that

(iv) $$E\ \| \int_E f\ d\ W\|^2 \leq K \int_E \| f(x)\|^2 \mu(dx) \quad \forall f \in L_E^2(\mu) = L_E^2(W)$$

Having established this much about the theory of vector valued stochastic integration, we proceed as follows:

Since E is of type 2 we can choose $K > 0$ so that (4.16) and (iv) holds. And since $f(x) = x$ belongs to $L_E^2(\mu) = L_E^2(W)$ we have that

$$U = \int_E f\ d\ W$$

has a centered gaussian distributed with covariance σ_X . In particular X is pregaussian. Now, let $\varepsilon > 0$ be given, then there exists a simple function f_o , such that (notice that $\int f\ d\mu = E\ X_1 = 0$)

(v) $$\int_E \|f_o(x)-x\|^2 \mu\ (dx) < \varepsilon^2 / K \quad \text{and} \quad \int_E f_o(x)\ \mu\ (dx) = 0 .$$

Put $Y_n = f_o(X_n)$, then $Y_1, Y_2, \ldots$ are independent, identically distributed with mean 0 and covariance

$$\sigma_Y(x',y') = \sigma_{f_o}(x',y') = \int_{E_o} \langle x',f_o(x)\rangle \langle y',f_o(x)\rangle \mu(dx)$$

Moreover, Y_n is E_o-valued for all n, where E_o is the finite dimensional subspace of E spanned by the (finitely many) values of f_o. Hence, by the finite dimensional central limit theorem we have (cf. (iii)):

(vi) $$V_n = \frac{1}{\sqrt{n}} \sum_{j=1}^{n} Y_j \overset{\sim}{\to} V = \int_E f_o(x)\, W(dx)$$

Let $\rho(x,y) = \|x-y\|$ be the norm-metric on E, we shall then compute the $\|\cdot\|_\rho$-distance between $\mathcal{E}(U_n)$ and $\mathcal{E}(U)$ (see 2.II.(6)). So let $\phi \in \mathrm{Lip}(E,\|\cdot\|)$ with $\|\phi\|_\rho \leq 1$, then we have

$$|E\phi(U_n) - E\phi(U)| \leq$$

$$\leq E|\phi(U_n) - \phi(V_n))| + |E\phi(V_n) - E\phi(V)| + E|\phi(U) - \phi(V)| \leq$$

$$\leq \|\mathcal{E}(V_n) - \mathcal{E}(V)\|_\rho + E\|U_n - V_n\| + E\|U-V\| \leq$$

$$\leq \|\mathcal{E}(V_n) - \mathcal{E}(V)\|_\rho + \{E\|U_n - V_n\|^2\}^{1/2} + \{E\|U-V\|^2\}^{1/2},$$

since $|\phi(x) - \phi(y)| \leq \|x-y\|$. By (4.16) and (v) we have

$$E\|U_n - V_n\|^2 = \frac{1}{n} E\|\sum_{1}^{n} (X_j - f_o(X_j))\|^2 \leq \frac{K}{n} \sum_{j=1}^{n} E\|X_j - f_o(X_j)\|^2$$
$$= K \int_E \|x - f_o(x)\|^2 \mu(dx) \leq \varepsilon^2 .$$

And by (iv) and (v) we have

$$E\|U-V\|^2 = E\|\int_E (x-f_o(x))\, W(dx)\|^2 \leq K \int_E \|x-f_o(x)\|^2 \mu(dx) \leq \varepsilon^2 .$$

Inserting this above, and taking sup over all ϕ with $\|\phi\|_\rho \leq 1$, we obtain

$$\|\mathcal{E}(U_n) - \mathcal{E}(U)\|_\rho \leq \|\mathcal{E}(V_n) - \mathcal{E}(V)\|_\rho + 2\varepsilon .$$

Thus by (vi) and (2.14) we obtain

$$\limsup_{n\to\infty} \|\mathcal{E}(U_n) - \mathcal{E}(U)\|_\rho \leq 2\varepsilon$$

and letting $\varepsilon \downarrow 0$, we see that $U_n \overset{\sim}{\to} U$ by (2.14). Thus the central limit theorem (6.1.4) holds.

Second proof of (6.1.3)⇒(6.1.4). Let X be an E-valued random vector, then we define:

$$c(X) = \sup_n E\,\| \frac{1}{\sqrt{n}} \sum_{j=1}^{n} X_j \| \quad ,$$

where $X_1, X_2, \dots$ are independent copies of X. And we put

$$CLT = \{ X \in L_E^0(P) \mid c(X) < \infty \}.$$

Then $(CTL, c(\cdot))$ is a Banach space, and DNA is a closed linear subspace of CLT. Actually DNA is $c(\cdot)$-closure of the set of all simple random vectors with mean 0 .

These propositions are by no means simple, even the fact DNA is a linear space is non-trivial. However, having established this we proceed as follows:

Since E is of type 2 we have (see(4.16))

$$c(X)^2 = \sup_n \{E\,\| \frac{1}{\sqrt{n}} \sum_1^n X_j \| \}^2 \leq \sup_n \frac{1}{n} E\,\| \sum_1^n X_j \|^2 \leq$$

$$\leq \sup_n \frac{K}{n} \sum_1^n E\,\|X_j\|^2 = K\, E\,\|X\|^2 \, ,$$

whenever $X \in L_E^2(P)$, $E X = 0$ and $X_1, X_2, \dots$ are independent copies of X . Now, let $X \in L_E^2(P)$ with $E X = 0$, then there exist simple random vectors Y_n with $E Y_n = 0$ and $E\,\| Y_n - X \|^2 \to 0$. But then $Y_n \in DNA$ and $c(Y_n - X) \to 0$. Hence, by the proposition above we have that $X \in DNA$, and so (6.1.4) holds. □

Theorem 6.1. gives us the answer "type 2" to all our questions. However, the Banach spaces, which naturally occure in probability, are rarely of type 2 (see Example 5.5). Thus it seems that Theorem 6.1., does not really help us. Of course, one could ask for less, e.g. one could ask: for which spaces E is the central limit theorem valid for pregaussian, mean 0 , finite second moment random vectors? The answer to this is :"either is E of type 2 or E is of cotype 2". Neither this helps much since cotype 2 is also rarely satisfied for the spaces arising in probability. However, the help is near, we only have to extend the notion of type to linear operators.

Let F and E be two Banach spaces and let T be a linear operator from F into E, then T is said to be of <u>type</u> p, if one of the following four mutually equivalent conditions holds (cf.(4.14)-(4.17)):

(6.4) $(T\,y_j) \in B_\varepsilon(E) \qquad \forall\, (y_j) \in \ell^p(F)$,

(6.5) $(T\,y_j) \in C_\varepsilon(E) \qquad \forall\, (x_j) \in \ell^p(F)$,

(6.6) $\exists K > 0 : E\,\|\sum_1^n \varepsilon_j\, T\, y_j\|^p \leq K \sum_1^n \|y_j\|^p \quad \forall\, y_1 \ldots y_n \in F$,

(6.7) $\exists K > 0 : E\,\|\sum_1^n T\, Y_j\|^p \leq K \sum_1^n E\,\|Y_j\|^p \quad \forall\, Y_1, \ldots, Y_n$ independent, mean 0, F-valued random vectors.

And exactly the same proof shows the following extension of Theorem 6.1:

THEOREM 6.2. <u>Let</u> E <u>and</u> F <u>be Banach spaces and</u> T <u>a continuous linear operator</u> : $F \to E$, <u>then the following four statemtns are mutually equivalent</u>

(6.2.1) $T\,Y$ <u>is pregaussian</u> $\forall\, Y \in L_F^2(P)$,

(6.2.2) $T\,Y$ <u>is pregaussian</u> $\forall\, Y$ <u>symmetric, discrete and satisfying</u> : $\|Y(\omega)\| = 1 \quad \forall\, \omega$,

(6.2.3) T <u>is of type 2</u>,

(6.2.4) $T\,Y \in \mathrm{DNA} \quad \forall\, Y \in L_F^2(P)$ <u>with</u> $E\,Y = 0$. □

Hence, if we want to show that an E-valued random vector X satisfies the central limit theorem (i.e. $X \in \mathrm{DNA}$), we just need to show that $X = T\,Y$ for some mean 0 F-valued random vector $Y \in L_F^2(P)$ and for some type 2 operator T from F into E. In applications F will often be a linear subspace of E (with a different norm) and T the injection : $F \to E$. Let us formalize this idea:

THEOREM 6.3. <u>Let</u> $(E, \|\cdot\|)$ <u>be a Banach space, and</u> $q : E \to \overline{R}_+$ <u>a lower semicontinuous seminorm satisfying</u>

(6.3.1) $\sum_{j=1}^\infty q(x_j)^2 < \infty \Rightarrow (x_j) \in B_\varepsilon(E)$.

<u>Let</u> X <u>be an</u> E-<u>valued random vector satisfying</u>

(6.3.2) $E\,X = 0$ <u>and</u> $E\,q(X)^2 < \infty$,

(6.3.3) $P(X \in E_o) = 1$ <u>for some</u> q-<u>separable set</u> $E_o \in \mathcal{B}(E)$.

<u>Then</u> X <u>belongs to the domain of normal attraction i.e. if</u> $X_1, X_2, \ldots$ <u>are independent copies of</u> X , <u>then</u>

(6.3.4) $\{ \frac{1}{\sqrt{n}} \sum_1^n X_j \}$ <u>converges in law in</u> $(E, \|\cdot\|)$ <u>to a centered gaussian measure on</u> E .

PROOF. Since $B_\varepsilon(E) \subseteq \ell^\infty(E)$, it follows easily from (6.3.1), that $\|x\| \leq K\, q(x)\ \forall x$, for some $K > 0$. Now, let $F = \{x \mid q(x) < \infty\}$, then by the lower semicontinuity of q we have that (F,q) is a Banach space, and the injection $T: F \to E$ is of type 2 by (6.3.1).

By (6.3.2) and (6.3.3) there exists an F-valued random variable Y , such that $E\, q(Y)^2 < \infty$ and $T\, Y = X$ a.s. Moreover, since T is continuous we have $T(E\, Y) = E\, T\, Y = 0$, and so $E\, Y = 0$ since T is injective. Thus the theorem follows from Theorem 6.2.□

EXAMPLE 6.4. Let $(S, \mathcal{B}, \mu)$ be a probability space, Θ a set, $(E_o, \|\cdot\|)$ a normed linear subset of R^Θ satisfying (2.73) , g a map : $S \times \Theta \to R$, and $Y_1, Y_2, \ldots$ independent S-valued random variables with $\mathcal{L}(Y_n) = \mu\ \forall n \geq 1$. And suppose that

(6.8) $G(s) = g(s,\cdot) \in E_o\ \ \forall s$ and $G \in L_E^2(P)$

(cf. Example 5.5.). Let m be the mean function of G :

(6.9) $m(\theta) = \int_S g(s,\theta)\, \mu(ds)$,

and let σ be the covariance of $H(s) = G(s) - m$:

(6.10) $\sigma(\theta, \lambda) = \int_S (g(s,\theta) - m(\theta))(g(s,\lambda) - m(\lambda))\, \mu(ds)$

Set $X_n = H(Y_n) = G(Y_n) - m$. Then $E\, X_n = 0$, $E\, \|X_n\|^2 < \infty$ and $X_1, X_2, \ldots$ are independent, identically distributed E-valued random vectors.

To prove that X_1 belongs to the domain of normal attraction, we need a lower semicontinuous seminorm q on E satisfying (6.3.1). I claim that (6.3.1) holds true if q satisfies the following two conditions

(6.11) $\exists K : \Theta \to R_+ : |f(\theta)| \leq K(\theta)\, q(f)\ \ \forall f \in E$,

(6.12) $\delta(\theta,\lambda) = \{ \sum_{j=1}^{\infty} (f_j(\theta) - f_j(\lambda))^2 \}^{1/2}$ is E-gaussian pseudometric, whenever $\{f_j\} \subseteq E$ and $\sum_1^{\infty} q(f_j)^2 < \infty$,

(see 2.VII(11)).

To see this let $\{f_j\} \subseteq E$ with $\Sigma_1^{\infty} q(f_j)^2 < \infty$. Then by (6.11) we have

$$\sum_{j=1}^{\infty} \| f_j(\theta) \|^2 \leq K(\theta) \sum_{j=1}^{\infty} q(f_j)^2 < \infty .$$

So, if (γ_j) is a sequence of independent real valued $N(0,1)$-distributed random variables, then

$$W(\theta) = \sum_{j=1}^{\infty} \gamma_j f_j(\theta)$$

converges a.s. ($C_\gamma(R) = \ell^2$) and W is a centered gaussian process with intrinsic metric (see 2.VII.(1))

$$\rho_W(\theta,\lambda) = \delta(\theta,\tau) = \{ \sum_{j=1}^{\infty} (f_j(\theta) - f_j(\tau))^2 \}^{1/2}$$

Now, let x'_θ be the linear functional: $< x'_\theta, f > = f(\theta)$, and let F be the linear span of $\{ x' \mid \theta \in \Theta \}$. Then $F \subseteq E'$ by (2.73) and F is a linear separating subset of E'. Now by (6.12) there exists a centered gaussian measure on E with intrinsic metric δ, and so by Theorem 4.1 (see also(4.2)), we have that $\Sigma_1^{\infty} \gamma_j f_j$ converges a.s. in $(E, \|\cdot\|)$. Thus (f_j) belongs to $C_\gamma(E)$ and so by (4.24), we have that $(f_j) \in C_\varepsilon(E)$. Thus (6.11) and (6.12) implies (6.3.1).

The conditions (6.3.2) and (6.3.3) take the following form in our case:

(6.13) $\int_S q(G(s))^2 \mu(ds) < \infty$,

(6.14) $\{G(s) \mid s \in S\}$ is q-separable.

Hence, if q is a lower semicontinuous seminorm on E, satisfying (6.11)-(6.14), then we have

(6.15) $\frac{1}{\sqrt{n}} \sum_{j=1}^{n} g(Y_j,\cdot) - m(\cdot)) \overset{\sim}{\to} W(\cdot)$ in $(E, \|\cdot\|)$, where W is an E-valued centered gaussian process with covariance function σ given by (6.10). Let us consider a special case:

1° Processes with continuous sample paths. Let Θ be a compact metrizable space and let $E = C(\Theta)$ with its sup-norm. Let ρ be a continuous pseudometric on Θ with the following property:

(6.16) Every centered gaussian process Y with $\rho_Y \leq \rho$, has a version with continuous sample paths.

And let $q = \|\cdot\|_\rho$ be the Lipschitz' norm introduced in 2.II.(9). Then clearly (6.11) holds, and if $\Sigma \|f_j\|_\rho^2 = K < \infty$ then

$$\{ \sum_{j=1}^{\infty} (f_j(\theta) - f_j(\tau))^2 \}^{1/2} \leq \sqrt{K}\, \rho(\theta, \tau) ,$$

so (6.16) implies (6.12). Then if g satisfies

(6.17) $$| g(s,\theta) | \leq M(s) \qquad \forall s \ \forall \theta ,$$

(6.18) $$|g(s,\theta) - g(s,\tau)| \leq M(s)\, \rho(\theta, \tau) \qquad \forall s \ \forall \theta, \tau ,$$

(6.19) $$\int_S M(s)^2 \mu(ds) < \infty ,$$

(6.20) $$\{ g(s,\cdot) \mid s \in S \} \text{ is } \|\cdot\|_\rho\text{-separable} ,$$

then (6.14) and (6.15) hold true and there exists a centered gaussian process W with continuous sample paths and covariance σ given by (6.10), such that

(6.21) $$\frac{1}{\sqrt{n}} \sum_{j=1}^{n} (g(Y_j, \theta) - m(\theta)) \overset{\sim}{\to} W(\theta) \quad \text{in} \quad (C(\Theta), \|\cdot\|_\infty) .$$

A continuous pseudometric satisfying (6.16) is called sub-C-gaussian. Let me give some examples of sub-C-gaussian pseudometrics on a compact metric space Θ. Let ρ be a continuous pseudometric on the compact metrizable space Θ, then in each of the follows three cases we have, that ρ is sub-C-gaussian:

(6.22) ρ is C-gaussian (see [19]),

(6.23) $$\lim_{\varepsilon \to 0} \sup_{\theta} \int_0^{\varepsilon} \sqrt{|\log \lambda (B_\rho(\theta, r))|}\, dr = 0$$ for some finite Borel measure λ on Θ (see [13]).

Here $B_\rho(\theta, r)$ is the open ρ-ball with center at θ and radius $r \geq 0$, i.e. $B_\rho(\theta,r) = \{\tau \mid \rho(\theta,\tau) < r\}$.

(6.24) $$\int_0^1 \sqrt{H_\rho(r)}\, dr < \infty \qquad \text{(see [15])}.$$

Here $H_\rho(\cdot)$ <u>is the entropy</u> of ρ, i.e.

(6.25) $$H_\rho(r) = \log N_\rho(r) \;,$$

where $N_\rho(r)$ is the covering number of ρ, i.e.

(6.26) $N_\rho(r)$ = the minimal number of ρ-balls of radius r, which cover Θ.

The conditions (6.17)-(6.19) are fairly innocent, but again the separability condition (6.20) makes troubles. Evan though $C(\Theta)$ is $\|\cdot\|_\infty$-separable (we have conveniently assumed, that Θ is compact and metrizable), the space $Lip(\Theta,\rho)$ is <u>not</u> $\|\cdot\|_\rho$-separable in general. It is an open problem whether (6.20) is always redundant, i.e. whether (6.16)-(6.19) alone imply the central limit theorem (6.21). In each of the cases (6.22)-(6.24) this can be shown to be the case (see [4],[13],[15] and [22]), but the general case is still open.

Let me finally remark, that the methods in [14] show that the results above carry over to the case where Θ is a polish space, E= $C(\Theta)$ with the topology of uniform convergence on compact sets, and (6.16)-(6.20) holds on any compact subset of Θ (cf. Example 5.5.2°).

2° <u>Empirical distribution function</u>. Consider the case described in Example 5.5.5°. In this case we have separability problems and the general central limit theorem seems not to apply. However, it can be shown (see [10]) that we have

(6.27) $$\sqrt{n}(F_n(\cdot,x)-F(x)) \overset{\sim}{-} U(x) \quad \text{in} \quad (B(R), \|\cdot\|_\infty),$$

where U is a time changed Brownian bridge, more precisely U is a centered gaussian process with the covariance function:

$$\sigma(x,y) = \min(F(x),F(y)) - F(x)\cdot F(y) \qquad \forall x,y \in R \;.$$

Or, if W is a standard Brownian motion on R_+, then

$$U(x) = W(F(x)) - F(x)\, W(1) \qquad \forall x \in R \;.$$

Again, it would be interesting to have an abstract vector valued central limit theorem, which would contain (6.27) as a special case.

REFERENCES:

The seven books below may serve as a basic reference to the subject. In particular, Kahane's exquisite book [5] is an excellent introduction (even though it is somewhat outdated by the six others), and it contains a wealth of inspiring applications.

[1] A.Araujo and E.Giné, The central limits theorem for real and Banach space random variables, Wiley and Sons, New York, 1980

[2] A.Badrikan et al.. Ecole d'eté de probabilités de Saint Flour V-1975, Springer LNS 539, Springer Verlag, Berlin 1976

[3] A.Beck (ed.), Probability in Banach spaces, Proceedings Oberwolfach 1975, Springer LNS 526, Springer Verlag, Berlin, 1976

[4] J.Hoffmann-Jørgensen et al., Ecole d'eté de probabilitiés de Saint Flour, VI-1976, Springer LNS 598, Springer Verlag, Berlin, 1977

[5] J.-P.Kahane, Some random series of functions, D.C.Heath and Co., Lexington Mass. 1968

[6] J.Kuelbs (ed.), Probability in Banach spaces, Adv.Prob. Vol 4, Marcel Dekker Inc., New York, 1978

[7] W.A.Woyczinski, K.Urbanik, Z.Ciesielski (ed.), Probability-Winter school, Proceedings 1975, Springer LNS 472, Springer Verlag, Berlin, 1975

The references below are completely random and scarce, and I refer to [1] - [7] for a complete list of relevant references.

[8] S.Bessaga and A.Pelsczynski, On basis and unconditional convergence of series in a Banach space, Stud. Math. 17(1958), p.151-164

[9] J.P.R.Christensen, Topology and Borel structure, North Holland, Amsterdam, 1974

[10] R.M.Dudley, Probabilities and metrics, Aarhus Univ.Mat.Inst., Lecture Notes Series No. 45, June 1976

[11] N.Dunford and J.T.Schwartz, Linear Operators, Vol I, Interscience Publ. Inc. New York, 1958

[12] U.Haagerup, Les meilleur constantes de l'inégalité de Khinchine, C.R.Acad.Paris 286(1978), p.259-262

[13] B.Heinkel, Mesures majorante et théoreme de la limite centrale dans C(S), Zeitschr.Wahrsch.u.verw.Geb., 38 (1977), p.339-351

[14] J.Hoffmann-Jørgensen, Stochastic processes on polish spaces, (to appear)

[15] N.C.Jain and M.B.Marcus, Central limit theorems for C(S)-valued random variables, J.Func. Anal. 19(1975), p.216-231

[16] S.Kwapien, On Banach spaces not containing c_o, Stud.Math.52(1974), p.159-186

[17] G.Köthe, Topological vector spaces, Vol I, Springer Verlag, Berlin, 1969

[18] M.Loeve, Probability theory, (Third ed.) Van Nostrand Co.Inc. Princeton, N.J. 1963

[19] M.B.Marcus and L.H.Shepp, Sample behavior of gaussian processes, Proceeding 6. Berkeley Symp. Vol.2(1972), p.423-441

[20] B.Maurey and G.Pisier, Series de v.a. vectorielles indépendentes et propriétés géometrique des espaces de Banach, Stud.Math. 58(1976), p.687-690

[21] F.Topsøe, On the Glivenko-Cantelli theorem, Zeitschr. Wahrsch. u. verw. Geb.,14(1970),p.239-250

[22] J.Zinn, A note on the central limit theorem in Banach spaces, Ann. Prob.,5(1977),p.283-286

COMPLETIONS AND THE NULL-COMPLETION OF VECTOR MEASURES

Davor Butković

We consider some relations between various completions of vector-valued measures and of measures which are in a natural way connected with them. This kind of questions is of interest in the theory of integration with respect to these measures (see [1] and [2]). In this note some earlier results by the author are improved, in particular the results on quasi-measures with values in locally convex spaces.

In all that follows X denotes a Hausdorff topological vector space, $\mathcal{A}$ denotes an algebra of subsets of some set $T \in \mathcal{A}$, and m denotes an additive set function on $\mathcal{A}$ with values in X. By $|\cdot|_{i\in I}$ we denote any family of F-seminorms (i.e. applications $|\cdot|_i : X \to R_+$ such that $|x+y|_i \leq |x|_i + |y|_i$, $|rx|_i \leq |x|_i$, if $|r| \leq 1$, and $|rx|_i \to 0$ if $r \to 0$ for $x,y \in X$ and a scalar r) which determines the topology of X. The completion of X will be denoted by $\hat{X}$, X_i denotes $(X/\{x:|x|_i = 0\})\hat{}$, and π_i denotes the quotient map $\hat{X} \to X_i$.

In the above setting there are at least two procedures to <u>complete</u> m, i.e. to extend m to a set function whose domain contains all null-sets of the extension (A is an m-null-set, $A \in N(m)$, if $m(S) = 0$ for every $S \in \mathcal{A}$ contained in A). The first procedure is to extend m to $\bar{m} : \bar{\mathcal{A}}^m \to X$, where $A \in \bar{\mathcal{A}}^m$ if there exist $L, M \in \mathcal{A}$ such that $L \subset A \subset M$ and $M \setminus L \in N(m)$; $\bar{m}$ is defined by $\bar{m}(A) = m(L)$ (the <u>Lebesgue</u> or the <u>null-completion</u> of m). The second extension $\hat{m}: \hat{\mathcal{A}}^m \to \hat{X}$ is obtained in the following way: $\hat{\mathcal{A}}^m$ is the algebra of all A such that for every neighbourhood V of zero in X there exist $L, M \in \mathcal{A}$, $L \subset A \subset M$, with the property that $m(S) \in V$ for any $S \subset M \setminus L$, $S \in \mathcal{A}$. $\hat{m}$ is then defined by $\hat{m}(A) = \lim_L \{m(L): A \supset L \in \mathcal{A}\}$ where the index set is directed by inclusion (the <u>Lipecki's Peano-Jordan type completion</u> of m, see [5] and [6]). Obviously, $\bar{\mathcal{A}}^m$ is contained in $\hat{\mathcal{A}}^m$ and [5, Ex.1, p.21] shows that the inclusion can be proper even for a scalar set function m. It is also obvious that $\bar{m}$ is the restriction of $\hat{m}$ on $\bar{\mathcal{A}}^m$. m reduces to $\bar{m}$ e.g. in the case of a metrizable X and $\mathcal{A}$ being a σ-algebra [5, Prop.2, p.21].

For every $i \in I$, the set function $\pi_i \circ m$ takes values in a metrizable space, and therefore is definitely more simple than m. What concerns completions of m and $\pi_i \circ m$, it is easy to check that $\hat{A}^m = \bigcap_{i \in I} \hat{A}^{\pi_i \circ m}$, regardless of the family I we choose (see[6 ,Th.2]). On the contrary the following example (also due to Z.Lipecki) shows that in general $\bigcap_{i \in I} \bar{A}^{\pi_i \circ m}$ depends on I :

EXAMPLE 1. Let $T = N$, let A be the algebra of finite and cofinite subsets of N, and let X be the space R^N of all sequences $(x_n)_{n \in N}$ with the topology of pointwise convergence. Let $m(A) = 1_A$. The topology of X can be generated by a family of seminorms $|(x_n)|_p = |x_p|$, $p \in N$, or by one F-norm

$$|(x_n)| = \sum_{n=1}^{\infty} \frac{|x_n|}{2^n(1+|x_n|)} .$$

In the former case $X_p = R$, $(\pi_p \circ m)(A) = 1_A(p)$, $N \setminus \{p\} \in N(\pi_p \circ m)$, $\bar{A}^{\pi_p \circ m} = P(N)$ (the power set of N), and $\bigcap_{p \in N} \bar{A}^{\pi_p \circ m} = P(N)$. In the later case $\bigcap_{i \in I} \bar{A}^{\pi_i \circ m}$ coincides with A.

In the most important case of a set function on a σ-algebra, $\bigcap_{i \in I} \bar{A}^{\pi_i \circ m}$ does not depend on I : in such a case $(\pi_i \circ m) = \overline{\pi_i \circ m}$ and $\bigcap_{i \in I} \bar{A}^{\pi_i \circ m} = \hat{A}^m$. It is to be remarked that in [2] the completion $\hat{m} \mid \bigcap_{i \in I} \bar{A}^{\pi_i \circ m}$ (denoted there by $\bar{m}$) is obtained by an alternative construction.

In our example m is a <u>quasi-measure</u>, an additive and exhaustive ($m(A_n) \to 0$ for pairwise disjoint $A_n \in A$) set function on an algebra; a <u>measure</u> is a σ-additive function on a σ-algebra. Also, in the above example X is a locally convex space, and for locally convex space valued quasi-measures and families $|\cdot|_{p \in P}$ of seminorms, $\bigcap_{p \in P} \bar{A}^{\pi_p \circ m}$ behaves similarly to $\bigcap_{i \in I} \hat{A}^{\pi_i \circ m}$ in the general case (see the theorem below). Denote the topological dual of a locally convex space X by X': then, for any X-valued set function m, $\breve{A}^m$ denotes the algebra $\bigcap_{x' \in X'} \bar{A}^{x' \circ m}$.

THEOREM. Let X be a locally convex Hausdorff vector space, let $m: \mathcal{A} \to X$ be a quasi-measure, and let $|\cdot|_{i \in I}$, resp. $|\cdot|_{p \in P}$, denote any family of F-seminorms, resp. seminorms, determining the topology of X. Then

$$\hat{\mathcal{A}}^m = \bigcap_{i \in I} \hat{\mathcal{A}}^{\pi_i \circ m} = \bigcap_{x' \in X'} \hat{\mathcal{A}}^{x' \circ m} ;$$

$$\bar{\mathcal{A}}^m = \bigcap_{p \in P} \bar{\mathcal{A}}^{\pi_p \circ m} \quad \Big(= \bigcap_{x' \in X'} \bar{\mathcal{A}}^{x' \circ m}\Big) .$$

Also,

$$\{\hat{m}(A)\} = \bigcap_{i \in I} \pi_i^{-1}((\pi_i \circ m)^\wedge (A)) =$$

$$= \bigcap_{x' \in X'} x'^{-1}((x' \circ m)^\wedge (A)) \qquad \text{for } A \in \hat{\mathcal{A}}^m ;$$

$$\{\hat{m}(A)\} = \bigcap_{p \in P} \pi_i^{-1}((\overline{\pi_i \circ m})(A)) =$$

$$= \bigcap_{x' \in X'} x'^{-1}((\overline{x' \circ m})(A)) \qquad \text{for } A \in \bar{\mathcal{A}}^m .$$

PROOF. First, let X be a normed space with the norm $|\cdot|$, and $m: \mathcal{A} \to X$ a quasi-measure. By [3,4.10,p.222], there exists $x_o' \in X'$ (a "Rybakov functional") such that $|x_o' \circ m|$ is a "control measure" for m ($\nu: \mathcal{A} \to R_+$ is a control measure for m if $\nu(A_n) \to 0 \iff \sup\{m(B): B \subset A_n, B \in \mathcal{A}\} \to 0$, $A_n \in \mathcal{A}$). It is easy to see that for a control measure ν for m, $\hat{\mathcal{A}}^m = \hat{\mathcal{A}}^\nu$. Since $|x_o' \circ m|$ is obviously a control measure for $x_o' \circ m$, $\hat{\mathcal{A}}^m = \hat{\mathcal{A}}^{x_o' \circ m}$. Similarly, by $\mathcal{N}(m) = \mathcal{N}(\nu) = \mathcal{N}(x_o' \circ m)$ we have $\bar{\mathcal{A}}^m = \bar{\mathcal{A}}^{x_o' \circ m}$.

Now, let us turn to the general case. Because $\bigcap_{i \in I} \hat{\mathcal{A}}^{\pi_i \circ m}$ does not depend on I, we have $\hat{\mathcal{A}}^m = \bigcap_{p \in P} \hat{\mathcal{A}}^{\pi_p \circ m}$. Let $x_p' \in X_p'$ be a Rybakov functional for $\pi_p \circ m$. Then $\hat{\mathcal{A}}^m = \bigcap_{p \in P} \hat{\mathcal{A}}^{x_p' \circ \pi_p \circ m}$, and as $x_p' \circ \pi_p \in X'$, $\hat{\mathcal{A}}^m$ contains $\bigcap_{x' \in X'} \hat{\mathcal{A}}^{x' \circ m}$. On the other hand, by the continuity of x', we have $A \in \hat{\mathcal{A}}^m \Rightarrow A \in \hat{\mathcal{A}}^{x' \circ m}$, and therefore $\hat{\mathcal{A}}^m$ is contained in $\bigcap_{x' \in X'} \hat{\mathcal{A}}^{x' \circ m}$ which gives the statement on $\hat{\mathcal{A}}^m$. The statement on $\hat{\mathcal{A}}^m$ is obtained analogously by $\bar{\mathcal{A}}^{\pi_p \circ m} = \bar{\mathcal{A}}^{x_p' \circ \pi_p \circ m}$.

Finally, $\pi_i(\hat{m}(A)) = (\pi_i \circ m)^\wedge (A)$ and $x'(\hat{m}(A)) = (x' \circ m)^\wedge (A)$ hold true on $\hat{\mathcal{A}}^m$ because $\pi_i \circ m$ and $x' \circ m$ are $\mathcal{A}$-tight additive extensions of $\pi_i \circ m$ and $x' \circ m$, and such extensions are unique [5, Th.2,p.23 and 6,Th.1]. The same argument holds for Lebesgue completions. □

Notice, that in general $\check{A}^m$ can be properly contained in $\hat{A}^m$, and that $\check{A}^m = \hat{A}^m$ in case A is a σ-algebra (then $\bar{A}^{x' \circ m} = \hat{A}^{x' \circ m}$ and $\bar{A}^{\pi_p \circ m} = \hat{A}^{\pi_p \circ m}$). Notice also that Example 1. above shows that $\check{A}^m$ can properly contain $\bigcap_{i \in I} \bar{A}^{\pi_i \circ m}$ (this was pointed out by the reviewer of [2] in Math.Rev. ; the argument for the opposite claim in [2] gives only $\check{A}^m = \bigcap_{p \in P} \bar{A}^{\pi_p \circ m}$). Observe, that in [1], $\hat{m}\,|\check{A}^m$ (denoted there by $\tilde{m}$) is obtained in an alternative way.

By the formulae in the theorem we have immediately the following

COROLLARY. For any locally convex space valued quasi-measure m the Lipecki's completion does not depend on the particular locally convex topology between the weak topology $\sigma(X,X')$ and the Mackey topology $\tau(X,X')$ in which m is exhaustive.

This corollary was stated in [1] for measures. The proof (with a few misprints), via the equality $\hat{A}^{\pi_p \circ m} = \bar{A}^{\pi_p \circ m}$, uses the assumption that A is a σ-algebra ; the result (and $\hat{m} = \tilde{m}$) are obtained by $\bar{A}^{\pi_p \circ m} = \check{A}^{\pi_p \circ m}$ and $\bigcap_{p \in P} \check{A}^{\pi_p \circ m} = \check{A}^m$. Observe, that m is a measure either for all the locally convex topologies between $\sigma(X,X')$ and $\tau(X,X')$, or for no one of them.

The next example (constructed for other purposes by Hoffmann-Jørgensen) shows that Theorem, and in particular $\hat{A}^m = \bigcap_{x' \in X'} \hat{A}^{x' \circ m}$, fails to be true in general without the assumption of exhaustivity on the additive set function m. Because $\bigcap_{x' \in X'} \hat{A}^{x' \circ m}$ is the Lipecki extension of A, if we consider X endowed with the weak topology, we see that without exhaustivity Corollary fails to be true as well. Observe, that exhaustivity in the weak topology means boundedness of m, and this does not imply exhaustivity in stronger locally convex topologies.

EXAMPLE 2. (see [4, Example 7, p. 29]). Let $T = [0,1)$, and let $\mathcal{A}$ be the algebra generated by the intervals $[a,b)$, $0 \leq a < b \leq 1$. Let F_n be the function defined by

$$F_n(t) = \begin{cases} 2nt, & 0 \leq t \leq 1/2n \\ 2-2nt, & 1/2n \leq t \leq 1/n \\ 0, & 1/n \leq t < 1 \end{cases},$$

let $m_n([a,b)) = F_n(b) - F_n(a)$, and let $m(A) = (m_n(A))_{n \in N}$ for $A \in \mathcal{A}$. Considering m as a c_o-valued measure, it is not exhaustive, and $\{0\}$ is not in $\hat{\mathcal{A}}^m$. On the other hand, for any $x' = (\ell_n)_{n \in N} \in \ell_1 = c_o'$, $(x' \circ m)(A) = \sum_{n \in N} \ell_n m_n(A)$ is σ-additive and extendible to a regular Borel measure on $[0,1)$. Moreover, $x' \circ m$ is absolutely continuous with respect to the usual Lebesgue measure (restricted to Borel sets). Therefore, $|x' \circ m|([0,1/2n)) \to 0$ if $n \to \infty$. This we can check directly: $|x' \circ m|([0,1/2n)) \leq \sum_{k=1}^{n} |\ell_k| 2k/2n + {}$ $+ \sum_{k=n+1}^{\infty} |\ell_k| \cdot 2$, and the first term (as well as the second) tends to zero by the Kronecker's lemma

$$\sum_{n \in N} |\ell_n| < \infty \Rightarrow \lim_{n\to\infty} (1/n) \sum_{k=1}^{n} k |\ell_k| = 0 .$$

We have $\{0\} \in \hat{\mathcal{A}}^{x' \circ m}$ for every $x' \in c_o'$, and therefore $\hat{\mathcal{A}}^m \neq$ $\neq \bigcap_{x' \in X'} \hat{\mathcal{A}}^{x' \circ m}$.

REFERENCES

[1] D.Butković, On Borel and Radon vector measures, Glasnik Mat. Ser.III., 13(33)(1978), 255-270.

[2] D.Butković, On integration with respect to measures with values in arbitrary topological vector spaces, Glasnik Mat. Ser.III., 15(35)(1980), 33-40.

[3] L.Drewnowski, On control submeasures and measures, Studia Math. T.L.(1974), 203-224.

[4] J.Hoffmann-Jørgensen, Vector measures, Math.Scand.28(1971),5-32.

[5] Z.Lipecki, Extensions of additive set functions with values in a topological group, Bull.Acad.Polon.Sci.Sér.Sci. Math. Astronom.Phis. 22(1974), 19-27.

[6] Z.Lipecki, Completion of additive set functions with values in a uniform semigroup, Glasnik Mat.Ser.III., to appear.

ON SOME COUNTEREXAMPLES IN MEASURE THEORY

Aljoša Volčič

The aim of this paper is to show how a technique, introduced first by P.R.Halmos to give a counterexample to the Radon-Nikodym theorem ([3], 31.9), can be used to produce two other counterexamples in measure theory. We suppose that all the measures are complete.

A measurable set M is said to be a measurable cover of a set E, if $M \supset E$ and if for every measurable set $G \subset M-E$, we have $\mu(G)=0$. Such a cover is obviously μ-unique. In [3], 12.1 there is an example of a set in a measure space (X, A, μ), which does not admit a measurable cover. The measure of this example takes however only the values 0 and ∞. We want to see what happens in less pathological cases, namely when the measure μ is essential, i.e. when

$$\mu(E) = \sup \{\mu(F): F \subset E, F \in A, \ \mu(F) < \infty\} \quad \forall E \in A .$$

We need also the following definition: an essential measure is said to be localizable [4] (or Maharam [2]), if for each collection G of measurable sets, there is a measurable set S satisfying the following two conditions:

$$\mu(G-S) = 0 \qquad \forall G \in G ,$$

if $T \in A$ is such that $\mu(G-T) = 0$, $\forall G \in G$, then $\mu(S-T)=0$.

In other words μ is localizable, if the lattice $A/\sim$ is complete (we identify sets E_1 and E_2 such that $\mu(E_1-E_2)+\mu(E_2-E_1)=0$). We shall use the notation $S = \sup G$.

The following theorem holds:

THEOREM 1. If μ is localizable, every set $E \subset X$ has a measurable cover $M(E)$.

Consider the following collection of sets:

$$M(E) = \{F: F \in A, \forall M \in A: M \supset F \Rightarrow \mu(F-M) = 0 \} .$$

We claim that $\sup \mathcal{M}(E)=M(E)$. In fact it is easy to see that

$$F \in \mathcal{M}(E) \quad \text{iff} \quad \mu(F-M(E)) = 0 .$$

We shall now give an example of a set in an essential (and non localizable) measure space, which does not admit a measurable cover.

EXAMPLE 1. Suppose A and B are two sets such that $\aleph_0 < \alpha = \text{card } A < \text{card } B = \beta$ and let be $X = A \times B$. Let us call a set of the form $\{(a,b_0):a \in A\}$ a horizontal line and a set of the form $\{(a_0,b):b \in B\}$ a vertical line. Let $\mathcal{A}$ be the σ-algebra of the sets $E \subset X$ such that $L \cap E$ or $L-E$ is countable for each horizontal or vertical line. Let us define on $\mathcal{A}$ the following two measures:

$\sigma(E)$ = the number of points of A such that $\{(a_0,b):b \in B\} -E$ is countable,

$\rho(E)$ = the number of points of B such that $\{(a_0,b):a \in A\} -E$ is countable.

The measure $\mu = \sigma+\rho$ is essential.

Let us decompose A into two sets, A_1 and A_2, having the same cardinality. Put $H = A_1 \times B$. Suppose that H admits a measurable cover $M(H)$. Such a set should contain (with the exception of a countable set) any horizontal line L. Therefore the set $M(H)-H$ is of cardinality $\beta\cdot(\alpha-\aleph_0) = \beta$. On the other hand, for each vertical line $L = \{(a_0,b):b \in B\}$ with $a_0 \in A_2$ we have $\mu(M(H)\cap L) = 0$, therefore $M(H)-H$ is of cardinality $\alpha\cdot\aleph_0= \alpha$, a contradiction.

REMARKS. This measure space was first considered in [3], 31.9 as a counterexample to the Radon-Nixodym theorem: ρ is absolutely continuous with respect to μ, but there is no f such that $\rho(E) = \int_E f\, d\mu$. The same measure space was again considered in [6] to show that localizability is not preserved under sums of measures : σ and ρ are (even strictly!) localizable, but their sum is not localizable. By the way, let us point out that also the product of two localizable measure spaces need not be localizable [2]. The example 1 was already presented in [7].

The problem whether the converse of Theorem 1 is true remains open: do the essentiallity and the existence of the measurable cover for each set imply localizability? As the spaces of Example 1 (one for each pair of cardinals (α, β) such that $\aleph_0 < \alpha < \beta$) are the only known examples of essential non localizable spaces, a negative answer would require a completely new approach.

In some questions concerning a Daniell integral I , the following condition assumes a particular importance (see [1], [9] and [10]): there exists a strictly positive real valued function u which is Stone-measurable, i.e. $u \wedge f$ is I-integrable for any I-integrable function f . Such a function u is called a weak unit. This condition appears as a natural generalization of the famous Stone's condition, requiring that $u \equiv 1$. The main consequence of the existence of a weak unit is the following representation theorem [8]:

THEOREM 2. If u a weak unit, the integral $\bar{I}(\bar{f})$ defined on the family of functions $\bar{\Gamma}_1 = \{\bar{f} : \bar{f} = f/u\}$ as $\bar{I}(\bar{f}) = I(f), f \in L_1(I)$, is a Daniell integral satisfying the Stone's condition. Therefore, there exists a measure $\bar{\mu}$ such that $I(f) = \int f/u \, d\bar{\mu}$.

Of course it is important to know if there are Daniell integrals without a weak unit. The following example answers positively the question.

EXAMPLE 2. Let A and B be as in Example 1 and let (A_1, A_2) and (B_1, B_2) be decompositions of A and B , respectively, such that card A_1 = card A_2 and card B_1 = card B_2 . Let us consider the following functions (C, C_1 and C_2 denote countable sets):

$$f_{a_o}(a,b) = \begin{cases} \text{constant } k & \text{for } a = a_o \text{ and } b \in B-C \\ 0 & \text{otherwise} \end{cases}$$

$$f_{b_1}(a,b) = \begin{cases} \text{constant } k & \text{if } b = b_1 \in B_1 \text{ and } a \in A-C \\ 0 & \text{otherwise} \end{cases}$$

$$f_{b_2}(a,b) = \begin{cases} \text{constant } k & \text{if } b = b_2 \in B_2 \text{ and } a \in A_1-C_1 \\ \text{constant } 2k & \text{if } b = b_2 \in B_2 \text{ and } a \in A_2-C_2 \\ 0 & \text{otherwise .} \end{cases}$$

Put $I(f_{a_o}) = k$, $I(f_{b_1}) = k$, $I(f_{b_2}) = K$ and define I by linearity on the linear combinations of the given elementary functions; then extend I with the standard procedure [5] .

It is easy to check, that for any function $u(.,.)$ which is Stone-measurable, we have

$$u(a_o,b) = k(a_o) \qquad \forall b \in B-C \; ,$$
$$u(a,b_1) = k(b_1) \qquad \forall a \in A-C \; ,$$
$$u(a,b_2) = \begin{cases} k(b_2) & \forall a \in A_1-C_1 \\ 2k(b_2) & \forall a \in A_2-C_2 \end{cases} .$$

Suppose that u is Stone-measurable, real valued and strictly positive. Suppose also that there exist a constant $k>0$ and a decomposition (A',A'') of A such that card $A' = \alpha' > \aleph_0$ and card $A'' = \alpha'' > \aleph_0$ and such that

$$a' \in A' \Rightarrow k(a') \in k$$

$$a'' \in A'' \Rightarrow k(a'') > k \; .$$

Each horizontal line $L(b_1) = \{(a,b): b=b_1\}$ has a countable intersection either with $X' = A' \times B_1$ or with $X'' = A'' \times B_1$.

Let us define

$$B' = \{ b : b \in B_1, \text{ card } (L(b) \cap X') \leq \aleph_0 \} ,$$

$$B'' = \{ b : b \in B_1, \text{ card } (L(b) \cap X'') \leq \aleph_0 \} .$$

At least one of the two sets B', B'' is of cardinality β. Suppose that card $B' = \beta$.

On the set $A' \times B'$ the function $u(a,b)$ assumes values not smaller than k on a subset K, which is a union of β sets having cardinality α' (because of the behaviour of the restrictions of u on the horizontal lines $L(b_1)$). So card $K = \beta$. On the other hand, looking at the behaviour of the restrictions of u to the vertical lines $L(a')$, the same set K is the union of $\alpha' \leq \alpha$ countable sets and therefore card $K = \alpha' < \beta$. The same contradiction is attained, if we suppose that card $B'' = \beta$. In this case we look at the cardinality of the subset G of $A'' \times B''$ on which $u(a,b) < k$.

The contradiction derives from the fact that A' and A'' are of cardinality $> \aleph_0$ or, equivalently, that $k(a)$ is essentially non constant on A.

We have therefore proved that

$$u(a,b) = k \quad \forall a \in A-C \ , \ \forall b \in B_1 - C_1(a)$$

($C, C_1(a)$ countable sets).

But this conclusion is not in agreement with the behaviour of the function u in the upper part of X. In fact, in $A \times B_2$ $u(a,b) \neq k$ on a set F (more or less it is $A_2 \times B_2$) having cardinality β. On the other hand, because of the behaviour of the restrictions to vertical lines, the set F has to be a union of α countable sets.

The contradiction follows from the hypotheses that u is Stone-measurable, real valued and strictly positive.

REMARK. The basic ideas of the construction of Example 2 are from [8]. In that paper A was restricted to be $[0,1]$ and the functions f_{b_2} were defined in a different way.

REFERENCES

[1] R.Becker, Sur l'integrale de Daniell, preprint.

[2] D.H.Fremlin, Decomposable Measure Spaces, Z.Wahrs. verv. Gebiete 45 (1978), 159-167.

[3] P.R.Halmos, Measure Theory, Van Nostrand, 1950.

[4] I.E.Segal, Equivalences of Measure Spaces, Am.Jour. of Math. 73 (1961), 275-313.

[5] M.H.Stone, Notes on Integration, Proc. Nat.Acad.Sci. U.S.A. vol. XXXIV (1948), 336-342, 447-455, 483-490; vol. XXXV (1949), 50-58.

[6] A.Volčič, Teoremi di decomposizione per misure localizzabili, Rend. di Matem. Roma (2) vol. 6, serie VI (1973), 307-336.

[7] A.Volčič, Localizzabilità, semifinitezza e misure esterne, Rend. Ist. Matem. Univ. Trieste, vol.VI, fasc.II (1974), 178-197.

[8] A.Volčič, Un confronto tra l'integrale di Daniell-Stone e quello di Lebesgue, Rend.Circolo Mat. Palermo ser. II t. XXVII (1978), 327-336.

[9] A.Volčič, Sulla differenziazione degli integrali di Daniell-Stone Rend. Sem.Mat. Padova vol. LXI (1978), 251-258.

[10] A.Volčič, Liftings for Daniell Integrals, to appear in the Proceedings of the Oberwolfach Conference on Measure Theory (1981), Lecture Notes in Mathematics, Springer-Verlag.

